M000035301

OPTIMAL DISTANCE

OPTIMAL DISTANCE

A Divided Life

PART ONE

Joan Carol Lieberman

Copyright 2017 by Joan Carol Lieberman

All rights reserved. In accordance with the U.S. Copyright Act of 1976, the scanning, uploading, and electronic sharing of any part of this book without the permission of the author and publisher constitute unlawful piracy and theft of the author's intellectual property. If you would like to use material from the book (other than for review purposes), prior written permission must be obtained by contacting Camperdown Elm Publishing, LLC, P.O. Box 39, Boulder, Colorado 80302 or: info@optimaldistance.com. Thank you for your support of the author's rights.

Publisher: Camperdown Elm Publishing, LLC, P.O. Box 39, Boulder, Co. 80302

First Edition: September 2017

The Bibliography and Biographical Index for *OPTIMAL DISTANCE, A Divided Life, Part One and Part Two*, can be found at the author's website, www.optimaldistance.com.

The names and identifying characteristics of a few individuals have been changed, whether so noted in the text or not.

ISBN 978-0-9987690-1-1

Library of Congress Control Number: 2017947063

Designed by T. Keith Harley

Pre-publication preparation by Scott S. Miller

Printed in the United States of America

CAMPERDOWN ELM PUBLISHING, LLC
P.O. BOX 39, BOULDER, CO 80302

TOPAZ MOUNTAIN – UTAH

circa 1890

Contents

For the Believers, the Unbelievers, and Those In Between

Salt Lake City—Der Mormonenstaat 1854

In 1854, my maternal ancestors were religious refugees struggling to survive in the desolation of the Valley of the Great Salt Lake, converts to beliefs so unacceptable their lives were in danger. My mother, a descendant of those Mormon pioneers, imagined God was speaking to her. My father, who was distantly related to the first and only Jewish governor of Utah, believed there was no God. Americans have always had beliefs as incompatible as those of my parents. We are divided by trust and suspicion, by veracity and deception, and by science and magical thinking. Yet, like my parents, despite disagreements, we have chosen to remain in close proximity, to lean into these dialogues and tensions. This book is dedicated to our continued coherence as we search for *optimal distance* from each other, from God, and from death—whether we are believers, nonbelievers, or, like me, in between.

About Optimal Distance

On Mother's Day 1981, I was two thousand miles away from my mother. The physical distance between us produced no longing, only vague relief. I was exempt from the Mother's Day tributes codified in American commerce because the mind of my sixty-two-year-old mother had no space for expressions of affection of any kind, not even from Hallmark. She had been a paranoid schizophrenic since my birth.

On that Sunday morning, I was a staying at the Mayflower Hotel in Washington, D.C., away from her and my Colorado home on a lengthy management consulting assignment. I walked up Connecticut Avenue in search of fresh air and coffee to Kramerbooks on Dupont Circle. While waiting to be seated, I bought *The Piggle—An Account of the Psychoanalytic Treatment of a Little Girl* by Donald Woods Winnicott, a British pediatrician turned psychoanalyst.*

The Piggle caught my eye because, having previously read some of Winnicott's extensive writings, his theories matched my own experiences as both a child and a mother. He had concluded that "good enough mothering" was sufficient to raise a healthy and happy child. With respect to mothering, Winnicott believed it was perfectly acceptable to earn a C+.

Winnicott saw the space between a mother and her infant, both the psychological and physical space, as the "holding environment." He believed that our capacity for happiness starts in that space, but with an essential caveat.

---

* Donald Woods Winnicott (born April 7, 1896—died January 28, 1971) was an English pediatrician and psychoanalyst best known for his ideas on the true self and false self, as well as the importance of transitional objects. He was the author of more than a dozen books and numerous papers, including: *The Piggle: An Account of the Psychoanalytic Treatment of a Little Girl.* London: Hogarth Press, 1971.

At the moment we take our first steps toward autonomy, away from the holding environment, thus changing the distance between ourselves and our mothers, we must believe that she will be there if needed.

I finished reading *The Piggle* as the sun was setting, finally understanding why psychological intimacy had always been hard for me, while my "precocious physical independence" had been essential to my survival. The seeds of this book began to germinate that night in my dreams. They reminded me that each of us is always searching for a comfortable distance from others. More importantly, what is comfortable psychologically and physically, continues to change throughout our lives.

All of us are born needing closeness to survive. As infants our brains literally shrink in the absence of physical contact. But if such closeness lasts too long, it can obstruct our emotional development. After a year or two, it is natural for us to begin to cautiously toddle away from our mothers, turning around frequently to make certain she is still close. Often we carry with us what Winnicott calls a "lovey"—a blankie, teddy bear, or other object of comfort. For the remainder of our lives, we are engaged in an endless dance of distance—always searching for more or less physical separation and psychological intimacy between ourselves and others.

It is rare for two human beings to maintain *optimal distance*, a perfect blend of psychological and physical intimacy, for an extended period of time. But our memories of such moments are powerful and have lasting impacts. A complicating factor is that our earliest experiences of *optimal distance* are soon intermixed with necessary efforts to socialize us. It is that stew of sensations which makes relationships with our mothers so tricky. Our first *optimal distance* experiences become mixed with feelings of vulnerability, dependency, and frustration, the inevitable by-products of being parented and growing up. We end up yearning for closeness while simultaneously resenting it.

This dance of distance begins with our mothers and fathers, or with a nanny, nurse, or foster parent. As toddlers we must continue the dance with our siblings and playmates. When we start school we are challenged to find a comfortable distance from our classmates and teachers. As adults moving into the world of work, we necessarily have to learn how to keep appropriate distance between ourselves and our supervisors, subordinates, and colleagues. Complaints about sexual harassment are examples of inadequate distance in workplaces and schools.

Most of us enter into one or more romantic relationships. The emotions produced by romantic love can be overwhelming, as if we are literally falling, giving us the same sensations of psychological and physical intimacy as between a mother and her infant. If we become parents, the dance begins all over again with our children.

My mother's earliest paranoid hallucinations began at my birth and kept me in harm's way. As a young child, I had no way of knowing that her behavior was being driven by the voices of invisible demons. Instead, I assumed something was wrong with me. My mother's facial expressions and body language told me that she did not love me. Because she was unpredictably violent, my nervous system began operating on high alert and never stopped.

I learned three basic lessons from those early years. First, my survival depended on staying away from my mother. Second, listening and observing were my only tools of defense. Third, it was essential for me to pay rent on my very existence with complete compliance to her rules and excessive generosity to others.

Growing up in Utah, my mother's hallucinations were hard for me to separate from the magical thinking of Mormonism. My mother's parents were the descendants of Mormon pioneers, but, unlike them, she stopped believing in Mormon doctrines during the Depression. With the onset of paranoid schizophrenia, she began hearing the voices of a god and a devil, who continued to whisper, or sometimes scream, in her ear until her death.

My father's parents were a Jew and a Catholic; something about that Gentile combination mixed with hardship turned my father into a committed atheist. My mixed religious lineage meant that in Utah I was born a full-blooded Gentile. Until I was fourteen, I always felt like a small wild animal desperately trying to hide from danger among a large herd of domineering dairy cows.

Throughout childhood I was convinced that no one else in the world had a mother like mine. Now I know that one out of every one hundred of us develops schizophrenia. It usually strikes almost overnight in the early years of adulthood. Males most often develop the disease when they are just finishing high school. The onset for females seems to come a few years later; in my mother's case, she was twenty-three. Even though it is a devastating disease in desperate need of attention, there are no schizophrenia-research marathons or gala balls for wealthy supporters. Nor do we see pastoral drug advertisements with split bathtubs on television calling our attention to this human disaster for which there is no cure.

Victims were once warehoused in insane asylums or large mental hospitals. Now, most people suffering from schizophrenia wander our streets and populate our prisons. A few, like my mother, are fortunate enough to have a loyal spouse, a parent, or a child who try to respond to the symptoms of schizophrenia without resorting to either lock and key or abandonment. Instead they spend their lives as unpaid adjuncts to our broken mental health system.

My mother became ill during the long era in psychiatric medical history when experts believed the disease was caused by "not-good-enough-mothering." In Western civilization, the archetypal blaming of women started with Adam and Eve, so it was too easy for a misogynistic cultural virus to infect the psychiatric community in the face of a mysterious untreatable disease. The viral vector was a 1934 study of forty-five schizophrenics, done without controls or peer review, whose authors concluded that schizophrenia was caused by a confusing mixture of maternal overprotection and rejection. Prominent psychiatrists and psychoanalysts subsequently coined pejorative labels for the mothers of schizophrenic patients, including "schizophrenogenic mothers" and "refrigerator mothers." Sadly, this practice of blaming mothers continued into the early 1980s.

I was almost forty when I read *The Piggle*. Despite several years of psychotherapy, I had not yet fully faced the implications of my early childhood. The most significant aspects of my history had either been kept secret from me or were stored in my unconscious. Winnicott's descriptions of his psychotherapy work with the little girl affectionately called "Piggle" caused me to recall more about how I had adapted to the circumstances of my birth. My father's presence was critical, as were emergency substitute mothers. Later, there was a teacher, a neighbor, and three friends whose presence in my life created healing holding environments. Winnicott reminded me why my lovey, a teddy bear named Teda, had been so critically important, as well as explaining why I was still drawing upon the remembered warmth and devotion of my first four pets.

As an only and often lonely child, my diary became an imaginary friend, a source of comfort and companionship at stressful times. Many of my diary entries, some no more than a few misspelled words written in childish script, became narrative signposts as I attempted to recreate my childhood path, sometimes wishing I was the Piggle talking to Dr. Winnicott.

By the time medical experts concluded that schizophrenia is a biological mental illness, my mother was dead. She had been born with a secret, a biological glitch in her brain most likely caused by a viral infection while in utero. Somehow the glitch was triggered in the early stages of adulthood at a moment of high stress or environmental toxicity.

My mother innocently carried her still secret schizophrenia into marriage and then into motherhood where it became our family secret. My father and I built walls, elaborate labyrinths, and underground shelters attempting to hide it. Schizophrenia is still a shameful secret for most families.

Secrets have enormous generative powers. There are no sterile secrets; each one reproduces. Some secrets breed immature desperate offspring. Others, like the one in my family, resemble a large beast who has wandered into a domesticated environment. We learned to quietly tiptoe past it—to modify our habits, our communications, and our dreams in order to protect ourselves from something wild and unpredictable in our midst.

Family secrets are our only guaranteed inheritances. Some are told, some are discovered, and others remain unknown. Just as my parents inherited the secrets of their mothers and fathers, I became the recipient of their unspoken bequests, some of which I have passed on to my progeny. I repeatedly poured sunlight and truth onto the secret separating my mother from the best parts of herself, as well as from all others. Despite my hard work, fragments have fallen into the cracks of my life and into those of my children. I have come to believe that each of us leave a legacy that is partially secret.

Yet secrets can also be helpful by allowing us to separate, to have our own space, to honor our uniqueness, and to preserve the deepest mysteries of our souls. Similarly, there is something in our biological makeup that seeks to protect us from the unbearable, hiding it from daily awareness. That defense mechanism allowed me to survive my childhood. Even though we see evil and experience terror, we may not remember having met it face-to-face. Instead, such encounters are locked away in our unconscious.

Neuroscientists believe we hold only fragments of sensory or visual memory from our first years of life, although parents sometimes provide their children with oral or pictorial snapshots of their early childhood. The reason so many adoptive children search so hard for their biological parents is our innate desire to know our earliest secrets. Of course, we cannot know everything about ourselves; nor do I mean to imply our histories are hidden because they are harmful. To the contrary, most are benign. If they have an emotional charge when revealed, it is because they resonate. Suddenly we understand why we prefer a certain food or, as twin studies demonstrate, why what we thought was our own unique quirk was really a genetic inheritance.

I was sixteen years old when I named our family secret in the kitchen of our California tract house. I hadn't planned my announcement; I simply blurted it out in a moment of adolescent desperation. It was like unbuckling a too-tight belt or taking off an emotional girdle. Breathing deeply, the sweat of shame evaporated. Then, I felt fearless. At our family table, I experienced internal cohesion for the first time. There was no space between what I saw and what I pretended not to notice. It was my first experience with my own kind of *optimal distance.*

* * *

BOOK I

CHILDHOOD

Among Jews, a Gentile refers to any person who is not Jewish.

For Catholics, a Gentile is anyone who is not a Christian.

For Mormons, a Gentile is anyone who is not Mormon.

In anthropology, of relating to, or indicating a nation or clan, especially a gens.

-1-

Inheritances of a Full-Blooded Gentile

I descended from four families who spent their lives in search of survival, trying to keep their religious beliefs a safe distance from others. Strung out across a great expanse of escapes, they were born with ready-made pedigrees of persecution as Jews in Spain, Alsace, and Germany, as Irish Catholics, and as a Presbyterian Scot, who switched his identity, converting to Mormonism in England, before sailing to America, and walking to the territory of Utah.

After generations of fight and flight, my mother and my father severed the cords to their religious bloodlines. Having turned away from their familial faiths, neither adhered to any tenet. Instead they became residents of separate psychological islands. The probability that they would ever meet, let alone marry, was extremely low. But as the fates would have it, they did, and as their only surviving child, I was born a full-blooded Gentile.

My mother was the youngest of five Mormon children. Her parents inherited that form of faith from their parents and grandparents, who were converted to Mormonism in England or Kentucky. As original Mormon pioneers, her ancestors arrived in the territory of Utah the hard way—most of them on foot, pulling handcarts, only a few fortunate females came in covered wagons. Her maternal great grandfather* was among the first group of Mormon pioneers to arrive in the valley of the Great Salt Lake with Brigham Young on July 24, 1847. He followed in his father's footsteps taking multiple wives during the era when there was no space or boundary between church and state in Utah—a place still rich in magical thinking.

The paternal ancestors of my mother were Scots. One son migrated to England, where he was converted to Mormonism before bringing his family to

---

*Her maternal great grandfather, William Cochran Adkinson Smoot, was the adopted son of Abraham Owen Smoot. His mother, Margaret Thompson McMeans, a divorcée, was the first of Abraham's five wives.

Utah via New Orleans. My mother's father, Edward Robert Beck, was the grandson of that convert. Born in Pleasant Grove, Utah, growing up muddy, reluctantly breaking Utah sod, he dreamed of becoming a gentleman, wanting to wear suits, clean crisp shirts, silk neckties, and polished Oxfords. He also wanted his hands to be free of blisters. He fell in love with automobiles and ended up selling the magic of motors to other Western dreamers, always in a suit, eloquently over-dressed.

Edward Robert Beck married Vermilla Smoot, the granddaughter of William Cochran Adkinson Smoot, in Blackfoot, Idaho on September 12, 1910. Their union was made under social duress because six months later, on March 1, 1911, Vermilla gave birth to their first child, a daughter, who was also named Vermilla. Eventually, the couple had another daughter and two sons before my mother, Margaret Audrey Beck, was born on June 7, 1919 in Salt Lake City. She was their fifth and last child.

Margaret Audrey Beck on her second birthday, Salt Lake City, June 7, 1921

As a toddler, Margaret began wearing a hat to protect herself from the attention drawn by her auburn hair. Nicknamed "Mugs" by her siblings and friends, Margaret attended South High School in Salt Lake during the Depression. Her wardrobe was so small that laundry was a part of her nightly homework.

Margaret Audrey Beck (center, front row) with three of her four siblings, parents, and maternal grandparents, Salt Lake City, 1932. *

Her body, like that of her father, was fully freckled. Despite her petite stature, she was hired to model nylon stockings for the Zion Cooperative Mercantile Incorporated,[1] then the most elegant department store between Denver and San Francisco. More importantly, many times "Mugs' ZCMI leg loot" represented the only cash in the Beck family's cupboard. Her sister Mary told me the Becks were so poor, they didn't notice the Depression.

By the time Margaret finished high school in 1936, her two older sisters had fledged and were busy building their own Mormon nests. Her mother took Margaret and her two brothers to Boise where her father was living alone as a pseudo-bachelor peddling Fords and Cadillacs. Margaret enrolled in Boise Business College, where she mastered shorthand and earned a degree in business.

---

* Left to right in front: Edward Robert Beck, Jack Campbell Beck, Margaret Audrey Beck, Vermilla Smoot Beck, William Cochran Adkinson Smoot. Rear: Edward Smoot Beck, Mary Ann Sarah (Polly) Jones Smoot, and Mary Isadore Beck

On September 1, 1939, the day the Nazis invaded Poland, the Beck family returned to Salt Lake City, moving into the Manx Apartments on State Street, a short block east of the Salt Lake Temple and the Hotel Utah. The day after Labor Day Margaret was hired by the Utah Power and Light Company, where proximity determined her destiny and mine.

Margaret was assigned to work next to Alice Liebermann, a slender, dark-eyed brunette who was not a Mormon. During breaks they began talking about the young men they were dating.

One day Alice said, "I've been meaning to ask, would you be willing to go on a blind date with my brother?" Describing him with pride, Alice added, "He is a federal research scientist. Right now he is stationed in Yuma, Arizona, but he often comes to Salt Lake on weekends to see me and our mother."

I wouldn't be writing this if Margaret had said, "Sorry, but I don't like blind dates." Instead she said, "I guess so."

Alice Liebermann, Ames, Iowa, 1938

Alice Liebermann's brother, Francis Valentine Liebermann, was born on August 26, 1911 in Newark, New Jersey. By the time he had his blind date with Margaret, he was known as Frank V. Lieberman, sometimes as F. V. Lieberman, but never as Francis Valentine, nomenclature he considered "too feminine." Frank was an unusual date for Margaret—he was eight years older and, like Alice, not a Mormon. He was a Gentile to Margaret, as she was to him.

Frank's father, Jacob Liebermann, was born to Jewish parents in New York; his mother, Anna Tuite, to Catholic parents in Newark—problematic religious lineage, particularly in Utah. On the upside, Frank V. Lieberman was well-educated with a doctoral degree in entomology and secure employment with the Federal government. He also was the owner of a 1939 Chrysler Imperial.

Frank V. Lieberman, Ames, Iowa, 1938

Cupid's arrow hit a bullseye. Frank was handsome and well-groomed. Margaret Audrey Beck was beautiful and charming. Over-ripe for marriage, Frank was instantly smitten. He proposed on their third date while they were having dinner at the Hotel Utah, a place of extreme extravagance, particularly for Margaret.

Margaret laughingly agreed to Frank's unexpected proposal with a flirtatious caveat, telling him she could not marry until she owned a black lace nightgown. The next day Frank purchased the most expensive black lace gown and peignoir in stock at ZCMI. According to Margaret's two older sisters, the cost of her prerequisite lingerie was one hundred dollars, then a ridiculous sum for the Beck family.

With respect to Margaret marrying a Gentile, Frank was so well-mannered, so well-educated, and so well-employed, it was likely impractical for her impoverished parents to object. Another issue was that Margaret had lost her faith in Mormonism during the Depression. Emphatic and desperate, she told her father that marrying Frank was her only chance never to be poor or hungry again. Expressing mild concern, Edward and Vermilla Beck allowed their youngest daughter to leave their Mormon clan—going from polygamy to exogamy in two generations.

Margaret Audrey Beck, Salt Lake City, 1940

But before giving her final approval, Vermilla Smoot Beck did what she did to all the suitors of her daughters. She served Frank V. Lieberman a biscuit covered with gravy on a too-small plate with lace porcelain trim. It was impossible for the biscuit to be eaten without gravy spilling through the lace trim. Did the suitor try to cover the stain with the plate? Pretend not to notice? Or did the young man forthrightly admit to the spill? Not only did Frank do the latter, he also demonstrated the best way to remove the stain, passing the Beck's "gravy on the tablecloth test" with flying colors. After that test of his character, Frank was invited to join the Beck family on a car trip to Yellowstone National Park in August 1940.

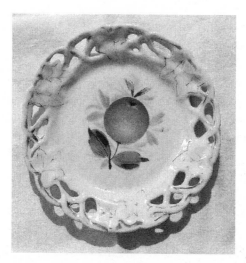

Beck's "gravy on the tablecloth test" plate

Margaret Audrey Beck and Frank V. Lieberman, Grand Canyon of the Yellowstone,
Yellowstone National Park, Wyoming, August 1940

Frank and Margaret wanted to be married at the Hotel Utah, the site of Frank's proposal, as well as the location of the marriage of their matchmaker. In February 1941, Alice Liebermann married Arthur William Marshall, the son of a Presbyterian dentist and a Canadian Methodist mother, in the Gold Room of the Hotel Utah. While the Mormon managers of the Hotel Utah did not object to hosting a union between two Gentiles, they declined to host a union between the daughter of Mormons and a Gentile. Somehow arrangements were made for Margaret and Frank to be married in St. Paul's Episcopal Church.

The marriage ceremony took place on March 27, 1941—the same day a coup in Yugoslavia overthrew the pro-Axis government. The bride and groom were each dressed simply. Frank wore a new navy gabardine suit; Margaret chose a wool crepe street dress in her favorite color, ice blue.

Unfortunately, the ceremony produced a close call. As Margaret started to walk down the aisle toward her groom, she caught sight of the Rector of St. Paul's, who was wearing a long white robe in the customary Episcopalian tradition. His appearance so frightened Margaret, she turned and ran out of the Church. Only after the Rector agreed to remove his robe, did the ceremony resume.

Following their interrupted vows, the bride and groom were honored at a homemade brunch at the home of Margaret's sister, Mary, near Liberty Park. The groom's divorced mother and the bride's parents and most of their siblings attended. Even though both families were Gentiles to each other, everyone was perfectly behaved. The nationalistic conflicts of World War II diminished distance due to social and doctrinal differences.

After brunch, the newlyweds escaped in a pre-arranged taxi to Temple Square, where Frank had hidden his prized 1939 Chrysler Imperial so it wouldn't be defaced with paint and trailed with cans by Margaret's two brothers. The groom turned west, driving along the shore of the Great Salt Lake, across the Bonneville Salt Flats to Wendover on the Nevada border, where they stopped to eat at the only coffee shop. When they reached Elko, the newlyweds checked into a motel where Margaret's father had made a reservation, requesting that the couple be given the so-called "honeymoon suite."

Many years later, Frank reported that his bride had donned the black lace nightgown and peignoir in the Elko Motel, albeit very briefly and only on that night. When the honeymoon was over, Margaret returned her premarital lingerie to ZCMI, telling the clerk the items were "not quite right for me."

Frank said he should have known then that something was wrong with his new bride, but, as an inexperienced husband, "I thought it was just how women were."

For the remainder of her life Margaret returned everything anyone gave her. Nothing was ever quite right.

Bride and Groom arriving at wedding brunch, Salt Lake City, March 27, 1941

-2-

A Female Prisoner of Two Wars

1941—1945

At the end of a two week honeymoon in California, Frank and Margaret began their married life in Delta, a small Mormon settlement in the middle of the Sevier Desert, one hundred thirty miles south of Salt Lake City. A few months before their marriage Frank had been deployed to Delta by the United States Department of Agriculture (USDA) to work with an experimental pesticide called DDT (dichloro diphenyl trichloroethane), studying the impacts DDT had on alfalfa weevils, as well as honeybees and other beneficial insect populations.[2]

The week before their wedding, Frank had moved from his single room in the Southern Hotel on Delta's Main Street into a small suite in the same establishment, one of the few rentals available. The Southern Hotel had numerous other tenants, so Frank had a part-time job keeping their small suite cockroach-free. After six weeks of hotel living, as Germany began bombing London, Margaret awoke in the middle of the night, convulsed with sneezes, her eyes swollen shut from an allergic reaction to Delta's spring pollens and perhaps also to the detritus of cockroaches.

At daybreak, his bride close to hysteria, Frank began driving above the speed limit toward Salt Lake City. Margaret lay on the back seat of the Chrysler, her face covered with a wet towel. Frank drove straight to the LDS Hospital. After a two-day stay, Margaret went home to her parents' apartment dosed up on an experimental drug that eventually came to be known as Benadryl.[3] Frank was once again a bachelor in Delta.

Margaret returned to Delta after the Fourth of July, but within hours not even the experimental drug could protect her from the outputs of rampant ragweed. She spent the rest of the summer of 1941 with her parents in her old bedroom. On weekends, Frank drove to Salt Lake to take Margaret into the canyons of the Wasatch Mountains in search of privacy and allergy relief.

After the first frost, Margaret returned to Delta a third time, only to suffer a severe asthma attack in late October. Frank once again rushed his bride to Salt Lake where doctors determined her asthma had been triggered by the hormones of pregnancy.

On December 1, 1941, Delta's newspaper, the *Millard County Chronicle*, reported, "Mr. and Mrs. Frank V. Lieberman have leased the Koiter home and are moving into it this week."

The Koiter's old frame house was surrounded by a white picket fence and a row of elm trees. On the day that Japan attacked Pearl Harbor, the newlyweds drove from Salt Lake to Delta, pulling a small trailer filled with what passed for Margaret's trousseau. Inside was an ice blue satin comforter (a wedding gift from her parents) along with the couple's first joint purchase—a new couch and matching chair upholstered in rust-colored damask velvet from Dinwoody's Furniture in Salt Lake City. The couch and chair were Margaret's choice—Frank had already learned to defer in matters of design.

Margaret's first full winter in Delta was as hard as the spring, summer, and fall had been. It wasn't only the environment—the howling winds, the desert sand mixed with desiccated snow, or the coyotes constantly prowling town in search of a loose chicken or a lonely cat—it was also her isolation. Frank was now responsible for a complex research project that kept him away from home for days at a time. He often traveled by tractor, hauling a sprayer rig filled with DDT—transport that required a slow pace on the shoulders of narrow two-lane roads, going as far south as St. George and as far east as Green River.

Margaret was one of the few married women in Delta without children; more notably, she was the only woman married to a Gentile. If you didn't go to meetings at the Mormon Ward House, the major points of contact were trade transactions at Thorton's Drugstore or Delmont's IGA Grocery Store. She tried to smile at the stares of others, but isolation made Margaret's already thin psychological skin transparent, so it became increasingly difficult for her to move through the small town gauntlet.

Whenever Frank returned from spraying, Margaret would beg him to take her to Salt Lake, reminding him that groceries were cheaper there. Pregnancy had given Margaret a craving for canned crabmeat, an item unavailable in Delta. Or she would feel desperate to see a new movie—one unlikely to ever be shown in the rodent-infested theater on Main Street. Some days she didn't know what she wanted to do except climb into the Chrysler and drive far away—the beginning of a coping pattern that persisted for the rest of her life.

Frank V. Lieberman spraying DDT, Millard County, Utah, July 1941

As the world continued to disintegrate in war, Frank and Margaret began baby preparations—shopping in Salt Lake for a crib and setting up a nursery in the spare bedroom. To treat Margaret's increasing loneliness and anxiety, they acquired a male springer spaniel, naming him "Army"—pet nomenclature drawn from the war-torn world.

From the moment Margaret learned she was pregnant, she was unusually outspoken about her preference for a boy. She had only episodic prenatal care because she so disliked being touched; her first pelvic exam was almost her last. In early May, Frank took Margaret back to Salt Lake to wait for the baby's arrival at the apartment of her parents. Only then did her doctor determine that Margaret was carrying twins.

On June 16, 1942, Margaret gave birth to twins, a boy and a girl. The son she had so wanted was stillborn, deformed,* and in the breech position—a complication requiring a Cesarean section to save the life of the female twin,

---

* Frank V. Lieberman reported that the deformation of his stillborn son was thought to be due to small maternal pelvis; his body was donated to science.

15

followed by a full hysterectomy to save her own. Not only had Margaret given birth to her first and only child, her hormones had been suddenly, prematurely, and permanently disrupted at the age of twenty-three.

-3-

Second Choice Survivor

Margaret Audrey Beck Lieberman holding the author, Delta, Utah, October 1942*

I am the surviving female twin. My birth as a twin and the complications of my mother's first and only pregnancy were two of many family secrets kept from me.

They named me Joan Carol. Frank wanted Joan; Margaret preferred Carol. Usage-wise her preference prevailed. After we were discharged from the hospital, Margaret's mother, "Nanie" as she was now called by all her children and grandchildren, cared for my mother and me in Salt Lake. The only good news was that Margaret's surgical menopause and changed hormonal status meant

---

* This is the only photograph of Margaret holding the author in the Lieberman family's photograph collection.

she was much less vulnerable to Delta allergens. When my father was anxious for his wife to return to Delta, Nanie told Frank he had to hire help.

"Frank, Margaret is not yet capable of caring for Carol. Whenever she is awake, she is crying most of the time, and what worries me most," Nanie whispered, "is that she seems unable to comfort her baby daughter!"

Frank, followed the guidance of his mother-in-law and hired June Davis, a young Mormon nursing student, to care for both his wife and child. Then living with her parents in Richfield, Sevier County, June Davis, eighteen, was the eldest of five children. Frank took June to Salt Lake where Nanie oriented her to the care needs of my mother and me before the return to Delta.

At about the same time, Marvin Wallway, his wife Bea, and their infant daughter, Mary Kay, arrived in Delta. Marvin was a young mining specialist sent by the War Department to search for sources of uranium needed for a secret bomb project. The Wallway family doubled the Gentile population of Delta overnight.

As the only two federal employees in Delta, Marvin and Frank shared office space. The Wallway family moved into a rental house on the opposite side of town, three blocks away from the Lieberman family. Bea, who was trained as a nurse, began providing emotional support to Margaret as she struggled to recover from the loss of her preferred son and her inability to bear more children, while June provided most of my care. The two young mothers shared war ration coupons, baked pies together, and talked as they pushed their baby daughters in prams up and down the wide dusty streets. At night, while their daughters slept, the two couples often played bridge together.

By September 1942, the Mormon population of Delta had many other Gentile strangers to worry about. Seemingly overnight, a camp of low wooden barracks was built just east of Delta. The opening of the Central Utah Relocation Center for eight thousand American citizens of Japanese descent made World War II real in a way that the death notices and heavily flagged military funerals for Millard County farm boys had not. Camp administrators soon realized the pronunciation of the original camp acronym sounded like "curse" so the name was changed to Topaz Relocation Center, after nearby Topaz Mountain.

The Federal Government put Frank in charge of the extermination of lice, bedbugs, and desert scorpions at Topaz. Because there were no skilled stenographers in Delta, Frank asked his boss in the USDA Washington headquarters for permission to hire a stenographer from among the Topaz internees.

His boss responded, "It's legal, but for God's sake, don't do that!" so Frank perfected his own hunt and peck typing system.

At nine months I started walking to keep up with Mary Kay, who was four months older and already on her feet. Army, the springer spaniel, who had learned to stay close by my side when I first began to crawl, took my walking efforts as an invitation to play and kept knocking me down. To distract Army, my parents acquired a female companion for him, naming her "Nurse." Both dogs playfully jumped on me, snuggled by my side, watching over me with unusual canine devotion, and soon produced a litter of loveable puppies.

The author (in hat) and Mary Kay Wallway, with neighborhood boys and puppies of Army and Nurse, back door of Koiter House, Delta, April 1944

Coincidentally, Bea became pregnant with twins in late August 1944. Dr. Myron Evans Bird, Delta's only physician, recommended bed rest. Initially, Margaret went to Bea's house for company. When the wind allowed, my mother would take Mary Kay and me for a walk, but without Bea for laughing and listening, Margaret found no pleasure in herding two toddlers.

My second birthday was the last day June Davis worked for my parents. She was engaged to be married and moved to Salt Lake while finishing her degree. Her disappearance from my life left a hole I didn't know how to fill.

Five months later in mid-November 1944, Margaret had a major psychotic break. The world was at war; no one could be trusted; anyone could be the enemy, and strike without warning. There was no safe place. The wind was a relentless reminder. The desert dust blew into the Koiter's creaking frame house through a thousand cracks. The radio seemed to short out in the middle of every song and newscast. Margaret thought the grocery store clerk looked at her suspiciously and took too long to count her war ration coupons.

Three rocky months followed. Then on the same day the U.S. Marines invaded Iwo Jima, Frank called Margaret to tell her that instead of coming home on Tuesday night, he would not return until Friday. His tractor had broken down and needed major repairs. The replacement parts were being shipped to St. George. Margaret would be alone for another four days.

A few hours later, the power went out. The sounds and scents triggered by the relentless desert wind caused Army and Nurse to bark incessantly. Opening her purse, Margaret discovered there were not enough gasoline coupons for her to retreat to Salt Lake. She felt trapped, as if the walls of the world were closing in on her. Margaret had been taken prisoner in her own home by an enemy only she could hear, but no one, not even she, could see.

The moon was almost full when the demonic voices awakened her just after midnight on February 20, 1945. Margaret was in bed, her back toward the bedroom door, when she heard it open. They entered and begin to call her name in high, whining voices. They threatened her with terrible destruction unless she did what they commanded.

> *"Marga ar r et!" Mar ga ar et! Your baby daughter is evil,"* they
> *screeched. "She is full of poisonous snakes! Kill the snakes before they*
> *strike! The snakes want to smother you! You won't be able to breathe!*
> *Remember when she first entered your body and stomped on your lungs?*
> *She will do it again! Be careful!"*

This was not the first time Margaret had been warned by the voices about the snakes. The first documented incident was November 1943;[*] from that moment on Margaret had become increasingly reluctant to approach her daughter. Often the voices went on for hours. Margaret would try to escape them by crawling into the bedroom closet, but the voices found her there, even though she was hiding in the dark. Margaret's tormentors made her promise not to tell anyone about them. No one could be trusted, not even her husband and certainly not Bea.

"No," the disembodied voices told her, "Bea is a bad woman! She flirts with Frank and makes better pies! Stay away from her! She is pregnant with more babies full of snakes!

When Bea laughed, Margaret heard the sound of the Devil. She stopped going to Bea's house just when her only friend most needed her. Bea couldn't fathom why Margaret had suddenly disappeared from her life, so she asked Marvin to ask Frank about the silence.

Frank's response was limited, "I don't know. She seems to be going through some kind of spell. Last week she told me the couch was full of poisonous snakes and only Carol could sit on it. I think it must be the loss of hormones or something. Tell Bea not to take it personally. I certainly don't."[†]

The voices commanded Margaret to send her daughter into the yard if she soiled her diapers. Margaret would scream until her daughter removed her diaper, struggling to put it in the soaking can. Other times Margaret would try to kill the snakes on the diapers—stabbing at them with sticks. Then on February 20, 1945, she tried to kill the poisonous snakes while they were still in her body.

---

[*] Alice Lieberman Marshall witnessed the admission of her niece to a Salt Lake Hospital that month. At the time Frank told his sister that Margaret had blamed a pothole in the street, saying: "The hole caused the pram to turn over." Admission was for broken arm and "infected diaper rash."

[†] Interview with Bea Wallway, Casper, Wyoming, August 1997

-4-

The Bear Goes to Topaz

Narrative constructed in 1992 from interviews and correspondence with three witnesses.

It was the second day the Father had been gone. The Child had been awake since dawn and knew not to move. There was no smell of pancakes. She carefully listened for the sounds of the Mother or the Bear, but the Child could only hear the trees outside her window creaking and the dust whirls moving tumbleweeds through the garden and down the road. The back screen door banged endlessly sending a jolt of electric-like fear through her body. Was the Mother or the Bear going to come today?

The Child knew the Wind made the Bear very dangerous and had learned how to play dead whenever the Bear appeared. She felt the cold wetness of her bed, the growl of hunger in her own stomach, and her mouth was dry from thirst, but she had learned that if she kept quiet, the Bear sometimes stayed away. She curled tightly into herself, pressing her fisted hands against her chest, and thought of the Dogs.

In the late morning the Child heard Music. The Mother was trying to make more noise than the Wind. The Music provided transport to another place. The Child drifted into sleep. When the Sun had reached the point where its warmth touched her back through the window, the door suddenly blew open. It was the Bear. It tore away her blanket and dragged her body out of the crib. Hooking a single paw into the collar of the Child's sleeper, the Bear pulled the Child's body behind it out of the nursery, through the living room, across the linoleum floor of the kitchen, out the banging screen door, and down the wooden steps into the yard. Instantly, the Dogs came from under the porch to the Child's side, threatening the Bear with nipping, barking, and growling.

The ground in February was half icy snow and half frozen grass, both now covered with a fine layer of desert sand from the daylong windstorm. The Child was face down on the ground as the Bear stood with a foot on her back. Its powerful paws broke a branch off a tree limb felled by the Wind from one of the elms surrounding the yard. The Wind and the Dogs howled in protest as the Child's soaked sleeper and soiled diaper were torn back and the Bear plunged the stick into the Child's anus. The Child screamed; the Wind and the Dogs howled louder. As the Child turned her head back to look at the Bear, the stick, half its length covered with bloody feces, was thrown at the dogs.

Instinctively the Child crawled away from the Bear. The Dogs followed, licking the Child's buttocks, ears, and her hands, tender offers of canine comfort. The Child listened carefully. There was the sound of a car door, a car motor, a car pulling away, and then suddenly stopping, backing up. Another car door sound and suddenly the Bear was back, coming nearer.

The Child began crawling again, trying to get away. But the Bear, seeing that the evil snakes were escaping, seized the coal shovel and, after first threatening the Dogs, began hitting the Evil Carrier of Snakes. The Child stopped, remembering to play dead. The Bear stopped swinging the shovel while the Dogs growled softly at a distance. The Bear then angrily dragged the Child's body, now cold, fouled, and stiff with terror and pain, toward the car.

The Bear raged, holding the Evil Carrier at arms-length. The rear door of the car was opened; the Child was put on the ground. A blanket from the trunk was roughly wrapped around her body to prevent the Bear from having further contact with the poisonous snakes. Then the Child was thrust face down onto the back seat and the car door shut. The Child heard the front car door opening and closing.

As the automobile began moving away, the Child could hear the Dogs following along the fence line in loud protest. The Child was mute, but opened her eyes and saw her Mother's handbag was on the floor below her. She worked one of her arms out of the blanket cocoon and took hold of the purse handle, holding it tightly.

The Bear drove fast and wild, growling at anything that got in its path, pounding the steering wheel, shouting at the Father, telling him that he would have to take care of the Child, he couldn't leave her alone again. The Child listened with acute sensitivity to everything the Bear said.

The Bear continued on the desert road until the guard station at the entry to the Topaz Relocation Center. The Father came here weekly to spread poison powder to kill the desert scorpions, lice, and bedbugs. Stopping at the camp perimeter, the Bear opened the back door to remove the Child, but then, as if distracted, suddenly turned and left the car in search of the Father. The Wind was turning the desert sand into stinging scattershot and the Bear disappeared into the furious cloud.

The Child noticed the rear car door had not latched. Still clutching the handle of her Mother's handbag, the Child backed her body across the seat of the car, moving like an inchworm in the blanket. Using her feet to push open the unlatched door, her body dropped to the ground and she blindly wormed her body away from the car. There was a dry ditch on the road's edge, the blanket fell away as her body rolled down the embankment into the ditch bottom. The Child, desperate to hide from the Bear, crawled into a dark culvert a few feet away. Her reserves spent, she slipped into unconsciousness.

When the Bear returned to the car and saw that the Child was gone, the Bear felt certain the guards at the front gate had been lying. The Father was at the camp, just as the Bear had thought! And the Father had come to the car and taken the Child, just as the Bear had wished! The Bear turned the car around and drove away.

Topaz Relocation Center near Delta, Utah, 1943

The next morning, one of the guard dogs caught the scent of the Child. The dog barked endlessly until a perimeter Guard searched for the source of the canine complaint. The Guard crawled into the ditch with a light to see for himself. Like the Bear, the Guard retrieved a blanket before he handled the

Child's body. He thought the Child was dead, until her eyes opened with the disturbance of being moved.

The Guard carried the Child into the mess hall and placed her on one of the tables. The Mother's handbag was opened. The Guard removed a book of war ration coupons and gave it to another guard, instructing him to take it to the camp administrator. While the ration book was being examined, the Child slowly woke to the warmth. The room began to fill with people. Women, men, and children, who looked alike, but who were different from the Mother and Father.

There was a strange smell. An old woman came to her side, gently lifted her head and put a warm cup to her lips. The Child opened her eyes to look. It was not the Bear. She drank. Then the safety of sleep came again.

When the Child next awakened, she was in a familiar place. Her body shook as she was being undressed. The Child kept her eyes closed until she was certain by touch, tone, and smell, that it was not the Bear. It was a Woman she knew and who knew her Mother. There was the scent of vanilla and she was wearing a red apron trimmed with white rickrack. The Woman's face was wet with tears and she was making music inside her mouth. The Child loosened her grip on the handle of her Mother's handbag.

The Child's sleeper was cut away and slowly pulled off her body where it had merged with raw skin. Lifting the Child into the sink, the Woman began pouring warm water over the Child, whose body shook. The Child could not feel the Woman's hand touching her back and buttocks, nor the soap and water that continued to pour over her. Later the Child's buttocks were covered with Unguentine ointment and a clean diaper and her badly cut thumb was bandaged. Mercurochrome was gently dabbed on the scalp and body wounds and zinc applied to the numerous insects bites. Nothing could be done for the bruises on her back. The Woman slowly edged the Child into a sleeper. The Child was offered warm milk in a tin cup and she drank. Then the Woman carried The Child to a cot near the kitchen stove and covered her with a blanket that smelled like the Wind and the Sun. The Child saw another child she knew nearby in what seemed like a dream.

A day and a night passed before the Mother arrived. The Child had made no sounds in the presence of the Woman and the other child, and she remained silent in the presence of the Mother. The Child's body was now flushed with fever and shook with chills. The Child held out the handbag.

Taking the handbag from the Child, the Mother asked, "Wherever did you find this?"

The Mother told the Woman that the Dogs had attacked the Child and she must have run away, and that she, the Mother, had been out searching for the Child for two days and nights.

In tears, the Woman silently nodded, before she gently helped the Child into the same car in which the Bear had taken the Child to Topaz and watched as the Child was driven away by the Mother.

When the Father returned on Friday night, the Mother told him the Dogs had knocked the Child off the porch, badly bruising her back. The Mother claimed the Dogs had bitten the Child on her hands and buttocks. Now the Child had a fever and no appetite. The Mother complained that the Child had a bad diaper rash due to the poor quality of the rationed laundry soap and that her raw skin had made the Child miserable to care for during the Father's absence. That night the Mother began insisting that the Dogs be destroyed because they could not be trusted around the Child. The Father, filled with disbelief, resisted.

Two days after the Father returned, the Woman gave birth to premature twins, a male and a female. The male twin lived one hour and twenty minutes. The female twin died about eleven hours later.

On March 1, 1945, one day after the Woman's premature twins were buried in the Delta City Cemetery, the Father reluctantly set out to follow the Mother's demand, taking the Dogs into a nearby field. The Dogs were joyful at an outing and sat obediently for their own execution. After the Father shot them with a borrowed hunting rifle, he buried them in the same field, digging their common grave with the coal shovel.

When the Father returned to the house, the Mother made him a breakfast of pancakes. Then the Father left for work. As soon as he was alone in his truck, he began weeping. He had loved the Dogs, had been delighted to see how devoted they were to the Child, and how much the Child loved them. The Father could not fathom why the Dogs had taken such a sudden turn in instinct. Only the Mother's testimony had allowed him to betray them.

By the time the Father reached Topaz, he was dry-eyed, having told himself that these were small losses in a world torn apart by war and unimaginable genocide. The Father slowly made his way into a place filled with visible injustice—a place where his work demanded targeted exterminations of a different nature.

Topaz Relocation Center at Night, 1945.

-5-

The Untouchable Sable Coat

I once overheard my father telling a friend that I had vomited any food my mother attempted to feed me near the end of World War II. He said Dr. Bird had recommended that items of finger food be left within my reach. When I was in sixth grade, I found a notation in my baby book, written in my father's hand.

> "Dr. Bird told us let her feed herself—the vomiting is just her attempt to catch your attention. Once you stop trying to feed her, the vomiting will cease. No child is capable of deliberate self-starvation."

The author with her father, Main Street, Delta, June 16, 1945

Many years later, my father told me, "The first food you ate with gusto after the deaths of Army and Nurse was on your third birthday. I took you down to Delta's Main Street and you gobbled down an ice cream cone I bought you as a birthday treat."

Lieberman and Wallway Families on the author's third birthday, Delta, Utah, June 16, 1945. Left to right: Marvin Wallway, Frank V. Lieberman holding Mary Kay. The author seated on chair; standing behind: left, Margaret Lieberman, arms crossed, and on the right, Bea Wallway.

I had no conscious memory of the day the Bear went to Topaz, nor of my hunger strike. Instead, my first mother memory consisted of two snapshots—both created within moments of each other on a sunny afternoon in September 1946 when I was four years old. One snapshot was captured in Kodak black and white; the other was preserved as a warning.

My mother and I were posing for my father on the deck of the *HMS Queen Mary*. As he looked down through the Kodak box camera lens, he motioned for me to move closer to my mother. She was wearing a new sable coat and a black velvet cloche, trimmed in copper-colored satin. I tentatively slid the palm of my hand across the intricately patterned sable pelts, downward, onto the silky softness of her calves encased in precious post-war nylons.

To me my mother seemed luminescent, glistening like the waters of New York's Hudson River in the afternoon sun. Her new sable coat, purchased that morning from a furrier on the lower East Side of Manhattan, was almost the same color as her hair. It was an ordinary moment of family pretense; yet, it is one of the few I remember in which my mother seemed magnificent—her mettle a match for any queen.

As soon as my father snapped the picture, I turned and looked up at my mother in admiration, accidently stepping on the toe of her black leather pump. At that moment, her countenance changed to a more familiar fierceness, an image missed by the camera, but perfectly preserved as one of my first conscious bytes of mother memory. It captured what I was learning the hard way: my mother was untouchable.

My parents and I were not embarking on a glamorous voyage across the Atlantic Ocean on the *HMS Queen Mary*. Instead, we were just three tourists from Utah seeing the sights of Manhattan. We had already reached our final destination, having traveled as far as we ever would as a family. The five weeks we spent crisscrossing the country in the Chrysler Imperial turned out to be the most intimate and sustained contact we ever had as a threesome. Now I see it as a sacred pilgrimage.

The author and her mother on the HMS Queen Mary, Pier Ninety, New York Harbor, September 1946

The trip was not the first or the last organized to distract my mother from her demons, but I doubt my father allowed himself to think of that as its purpose so early in their married life. Instead, at age thirty-five, he more likely felt a need to make his way back to his birthplace, Newark, New Jersey, stopping at other life landmarks, his college alma mater in Ames, Iowa, and then Chicago, where his two older brothers had built careers and families.

My mother, then twenty-seven, had never been further east than Utah. Yet, she arrived in Manhattan with all the attributes of a native. Stunningly stylish, her always-frugal husband had defied his Depression-shaped character by adorning her with sable and silk stockings. Naive observers might have mistaken my mother for the wife of John Jacob Astor. But Margaret Audrey Beck had become the wife of Frank V. Lieberman, and, as chance would have it, my father had been raised to be a perfect match for her invisible imperfections.

As we traveled together, a fateful shadow fell over our threesome, the distant effects of traumas that dated all the way back to 1919. For my father, 1919 was the year that the adulterous behavior of his father, Jacob Liebermann, ended the marriage of his parents in Ogden. To the majority population of Utah, my father's mother, Anna Tuite Liebermann, became a destitute Catholic Gentile abandoned by a Jew, without means of support for herself and her four young children.

As for my mother, in 1919 the epidemic of Spanish Influenza had already spread across the world causing the deaths of millions. In January 1919, Nanie, then five months pregnant with my mother, was stricken while caring for her sister Genevieve, and her ten-year-old niece, Beulah, Genevieve's only child.* Despite Nanie's desperate home nursing, Beulah died. While both Nanie and Genevieve survived, after many years of research, I now believe that the consequences of that deadly viral assault during the fifth month of my mother's fetal life were manifest on the day the Bear went to Topaz.

---

*Nanie's sister was Genevieve Smoot Spettigue. Her niece, Beulah Spettigue, born May 27, 1908, died of the Spanish Influenza on January 25, 1919 in Salt Lake City.

-6-

Ben's Birthday Couch

World War II and the War on Weevils ended about the same time. The USDA decided Frank V. Lieberman was the right man to be the Station Manager of USDA's Experimental Research Laboratory at Utah State University in Logan. The population of Logan in 1946 was about fourteen thousand. It was still a small town, but ten times bigger than Delta, with two movie theaters, and a non-Mormon population of almost one thousand.

Logan LDS Temple and Rocky Mountains, Cache Valley, Utah, 1947

Like most communities after World War II, rental units in Logan were in short supply. We moved into the home of a fifty-year-old Swedish bachelor, an attorney, who had a very large house all to himself. The boarding arrangements had been made by his brother, a Millard County farmer my father met while dusting his weevil-infested fields with DDT. In return for sheltering us, the Mormon bachelor obtained the benefits of my mother as a built-in housekeeper and cook. He quickly became infatuated with her, imagining a kind of reverse polygamy—fantasizing she would take him as a second husband. My untouchable mother found his attentions intolerable.

Desperate to escape her admirer, after three months, my mother placed an ad in the *Logan Herald Journal* offering a fifty dollar reward for information leading to any available rental, an illegal act under post-war regulations. My father was issued a summons, but not before a thrifty homeowner offered to rent us space in their dark basement. The summons was dismissed after local officials learned my father was in Logan on an official assignment for the federal government.

There were three rooms in our new basement home. The first became a tiny living room; the second became our single bedroom. The third room lacked even basement windows. Its centerpiece was a green Maytag wringer washer strategically located over a spidery drain. The four corners of this windowless area held the other essentials of independent living: a tiny electric stove, a small refrigerator, a toilet next to a large laundry sink, and a showerhead that drained into the center of the room. The toilet and shower were made semi-private by shower curtains hung from the joists of the floor above us.

In the middle of the wall between the toilet and the shower was a large coal-burning furnace. The Maytag room was so crowded we ate our meals on a card table in the tiny living room. Most of our furniture, including the snake-filled Dinwoody's couch, remained in storage.

While the move for my parents was a desperate descent into sub-standard shelter, for me it turned out to be heaven on earth because Marlene Evans lived next door. As Marlene stood on her front porch watching our move-in parade, I stood on the sidewalk staring at her, holding my lovey, Teda, a teddy bear. Eventually Marlene's mother, Afton, came out on the porch carrying a pitcher of Welch's grape juice and a plate of Wonder Bread peanut butter and jelly sandwiches cut into quarters. These were the sacraments used to anoint us as new members of the neighborhood.

Marlene was already attempting to teach me how to jump rope by the time her father came into view while walking home from work. Ben Evans was a gentle giant. Six feet, four inches tall and always soft-spoken, he moved as if he had never known fear. Since I was trying to hide my own terror, Ben Evans became my iconic model. I have been trying to demonstrate the fearless equanimity of Ben Evans all my life, never reaching anything close to his manifestation.

Marlene Evans, holding Teda, and the author, Logan, Utah, 1947

Ben owned and operated Ben Evans' Garage, but he made his way around Logan disguised as a banker in a suit and a brown felt hat. Arriving at his garage, Ben would strip down to his Mormon Temple garments,[4] in order to put on a heavy denim jump suit, before beginning to work his practical magic on clutches and carburetors. Outside his garage, Ben always appeared in banker-like attire. After he was elected Logan City Commissioner, it was sometimes necessary for him to change clothes several times a day in order to make the requisite appearances at meetings and roadbed inspections.

In contrast to her husband, Afton Lee Evans was a small woman. She registered four feet, ten inches on the kitchen door frame marked with the heights of her children. Those penciled notations documented the power of her husband's genes. By the time they were twelve, Afton's three older children had passed her frozen mark and Marlene was not far behind. Her short stature may have been why Afton bristled with the fierceness of a small cornered animal. She was still running a very tight housekeeping ship, but her interest in cooking was over. The main culinary resources in her kitchen were Cheerios, milk, Wonder Bread, peanut butter, Welch's grape juice and Welch's grape jelly.

Ben had made the necessary adaptation by taking a hot midday meal at the Blue Bird Café on Main Street. Afton often called Ben near the end of his work day instructing him to meet her and the children for supper at the Blue Bird

Café because she had been so busy with the Relief Society,[5] she had not found time to put a meal on the family table. Removing her omnipresent hairnet, Afton would change into her good Penaljo shoes and, with her four children in the backseat, drive four blocks to dinner in the family Buick.

In search of a safe distance from my mother, I slipped into the terrain of the Evans family as fast as I could. Marlene's friendship was one of the greatest gifts of my childhood. I admired everything about her. She lived in a whole house, shared a big bedroom with her older sister containing a large collection of neatly dressed dolls from foreign lands, standing or seated on shelves that also held many interesting books. Marlene could already read, write, draw, and play the piano—at age six, her unusually long fingers could even span a full octave.

I have no memory of Marlene resisting any form of parental authority, certainly not piano practice. Afton would call Marlene away from our play, turn on the piano light, and set the metronome before pulling up a chair to supervise. It took artful begging, but I finally persuaded Afton to let me to stay while Marlene practiced. I would lay on the carpet in their living room, quietly paging through one of Marlene's books, sometimes falling asleep, lulled by the sound of repetitive practice scales.

In the living room there was a single brown mohair couch where Ben napped on Saturday and Sunday afternoons because Afton's rule was that Ben was not allowed to lay down on the bed he shared with Afton after she had straightened the linens and pulled the chenille bedspread tightly across its surface. The same rule applied to all the other beds in the house. The brown mohair couch was too short for Ben to stretch out on. Something, probably a cricked neck, led Ben to announce that he wanted a new couch for his fifty-second birthday—a couch seven feet long.

Ben's birthday couch was the main topic of discussion in the Evans' household for weeks. A couch of that length would have to be custom made by a company in Salt Lake City. During spring vacation, Afton invited me to accompany Marlene and herself on a day-long couch shopping expedition.

We traveled by bus to Ogden because Afton didn't drive on highways. In Ogden, we climbed aboard the "Bamberger,"[6] a train running between Ogden and Salt Lake City. While we two girls wriggled around the showroom, Afton took a long time to make her decisions about style and fabric, confirming measurements, and delivery arrangements.

Eventually Afton finished and treated us to a late lunch of tuna salad sandwiches and strawberry sodas at ZCMI. Our trio climbed aboard the Bamberger train to Ogden; then all three of us fell asleep on the bus to Logan. When Marlene woke with motion sickness, Afton's fierceness persuaded the bus driver to stop just in time to allow Marlene to lose her special ZCMI lunch by the side of the road in Brigham Canyon.

Six weeks later, two burly men struggled to unload Ben's birthday couch which was completely covered with a protective plastic cover. Upholstered in pale satin brocade, Ben's birthday couch looked exceedingly grand. The ready-to-be-retired brown mohair couch was left in its regular spot on the south wall, while Ben's roll top desk was moved to the boys' back bedroom to make room for Ben's birthday couch on the north wall, where it instantly made everything else look shabby.

On Saturday afternoon, when I peeked through the front screen door, Ben was taking a nap on his birthday couch—the thick protective plastic cover lay in a heap on the floor. Ben seemed to be smiling in his sleep.

Later Marlene and I were playing in her backyard where we couldn't help but overhear Afton and Ben arguing in the kitchen about whether Ben could sleep on his birthday couch without the plastic cover. Ben was adamant that he could and he would. Afton was equally adamant that he shouldn't and he wouldn't. Listening to their escalating debate, Marlene and I took Ben's side. The plastic was icky—worse than the feel of the brown mohair on bare skin. Besides, the couch was Ben's birthday present.

Ben punished Afton for her stingy sense of comfort by taking his Sunday nap on the old brown mohair couch. For the next several weeks, Ben worked late and had supper alone at the Blue Bird Café. Afton and the children stayed home and ate Cheerios and milk for supper. As far as I know, no one sat on Ben's birthday couch, let alone took another nap on it. The birthday couch was slowly transformed into a memorial to the Evans' marital impasse where stacks of bills, automobile parts catalogues, and the minutes of Logan City Commissioners meetings accumulated. Once a week, Afton swiped her ostrich feather duster over the piles of paper stacked on top of the protective plastic cover.

Couches already held disproportionate importance for me given my mother's repeated declaration, warning of the poisonous snakes inside the cushions of our Dinwoody's couch. One of my earliest attempts to make sense of the world led me to wonder whether something about couches made women dangerous.

After I started piano lessons, even though our piano was still in storage, I sometimes practiced on Marlene's piano while she was at school and Afton was at Relief Society lessons. Whenever I was alone in the Evans' living room on the pretense of piano practicing, I would kneel down on the floor by Ben's birthday couch and carefully lift the protective plastic cover to put my cheek against the soft satin brocade. I was learning how to pray in Mormon Primary School, so one afternoon I uttered my first independent prayer, naively hopeful that God would help Ben.

"Heavenly Father," I whispered, "Please make Afton let Ben lay down on his birthday couch!"

-7-

Afton's Broom

Marlene's seventh birthday in mid-July loomed on the horizon and her energetic enthusiasm must have pressured my mother to do something for my fifth in mid-June. Marlene arranged for a Sunday afternoon party on the back lawn with Vicky Greaves and Alda Funk, girls who lived on the block. After pin the tail on the donkey, we opened ZCMI paper ball surprises and ate cake with ice cream on an improvised table. My presents were two coloring books with two boxes of crayons and a set of jacks. On Monday, a pink diary with a lock and key arrived in the mail from my Aunt Mary in Salt Lake City.

The day after the party, Marlene and I were coloring in my new books as we sprawled out on one of my father's surplus Army blankets in the Evans' backyard. Afton surprised us with a tray of peanut butter and grape jelly sandwiches and old jelly jars full of grape juice—our own impromptu picnic lunch. After eating, I needed to make a bowel movement; I told Marlene, who said she needed to do the same. We did not want to go into Marlene's house because it was so close to piano practice time that Afton might not let Marlene come out again. I did not want to go into my home for different, but equally compelling, reasons.

These circumstances led Marlene and me to turn the ground beneath the Evans' large lilac bush, then in glorious bloom, into an outhouse. It was hard to tell which bowel movement belonged to me and which one belonged to Marlene. The similarity was a source of special comfort to me, but I didn't tell Marlene. She would not have been surprised, since she was already well aware I wanted to be like her in every possible way.

I no longer remember who discovered our dual acts of defecation, but I do remember my mother scrubbing me all over with the big brush she usually used on the cement floor. Marlene and I were not allowed to play together for a week. During our penalty period, I would stand on my parents' bed and pull myself up to the window where I could watch Marlene skip rope and play hopscotch by herself on the front sidewalk. Each day of isolation was longer than the previous one. That week I learned what it was like to be a prisoner on home detention with a window looking out on life.

Punishments over, Marlene and I once again began spending all of our time together. On the Fourth of July, my parents went to a faculty picnic in Logan Canyon, but I successfully begged to stay home with Marlene. Because it was hot and a holiday, Afton allowed us to run through the sprinkler on the front lawn. Marlene had a bathing suit, but I didn't. Afton frowned as I stripped down to my underpants, but I quickly forgot her look of disapproval as we ran back and forth through the water screaming in delight until the sun began to fade.

Marlene went inside for towels because it was time for us to go down the block to eat hot dogs at Vicky Greaves' house. I sat down to rest on the grass beneath the big elm tree that grew in the parking strip in front of the Evans' house. Looking up at the pink clouds forming in the sky, I wanted darkness to come quickly so we could light our sparklers.

As I leaned back against the trunk of the elm tree to rest, something terrible began happening. Struggling to my feet, my body seemed to be melting, leaking. There was shameful moist movement all over my bare back. The tree trunk had been covered with camouflaged gypsy moth caterpillars! Urine ran down my legs onto the sidewalk; I opened my mouth to scream, but no sound came out.

Camouflaged gypsy moth caterpillars on elm tree.

As Marlene came out on the porch with the towels, she saw my face contorted by terror. She ran to my side, but immediately began backing away.

Safely on the other side of the sidewalk, Marlene started jumping up and down, while screaming, "Help her, Mother! Help her!"

Afton started to come toward me, then turned away, running in the opposite direction. Assuming abandonment, I lost all hope. But then Afton came running back toward me, holding her big broom up upside down up, over her head. She began sweeping the caterpillars from my back, while commanding Marlene to take the sprinkler attachment off the hose and screw on the nozzle. Then Afton stood a broom stick away from me, using the hose to wash off the remaining caterpillars.

Sternly instructing Marlene to hold a towel as a screen from the street, Afton helped me take off my underpants, stained with yellowish-green caterpillar entrails. I was devastated, naked, shaking with shame on the wet wormy walk.

Following a strawberry bubble bath in their tub, Afton dressed me in one of Marlene's old nightgowns. Then Marlene and I sat on the front steps watching Ben as he moved around in the dark with huge newspaper torches burning the gypsy moth caterpillars off the trunk and lower branches of the elm tree. After Ben lit our sparklers, Marlene and I ran around in tiny circles, keeping a safe distance from the tree. We never did go to Vicky's house for hot dogs. Instead, Afton served us a bowl of Cheerios with milk.

The next day my father arrived with his USDA tractor rig to spray all the trees on the block. I missed his efforts because Afton had taken Marlene and me to a matinee movie, Disney's *Make Mine Music* at the Capital Theater.

For the remainder of the summer and many months after, I carried my toy broom wherever I went, repeatedly reenacting Afton's rescue of me with her weapon of choice.

The author and her toy broom, with her father and Teda, Logan, September 1947

-8-

My Mormon Safe Havens

In many ways my life was saved by the Church of Jesus Christ of Latter Day Saints, just not in the ways promised by the Prophet Joseph Smith in his *Book of Mormon.* I quickly learned the Church of Jesus Christ of Latter Day Saints had a perfect protection plan for me; there is something important for Mormons to do most days of the week.

A few days after we moved next door to the Evans family, I followed Marlene to Primary School as part of my instinctual strategy to stay away from my mother. I felt certain she would never follow me based on her rigid refusals to attend any kind of Mormon meeting whenever we visited her family in Salt Lake. Also, she seemed to sleep a lot. (In retrospect, this was likely due to her daily intake of Benadryl.) My father, who was gone a great deal of the time, habitually deferred to my mother to avoid conflict.

Whenever I entered the Ward House,[7] I felt like a small wild animal trying hide out in a large herd of Mormon dairy cows. The price of admission to this place of safety required that I sit very still through Primary School classes, as well as the long Sacrament and Testimony Meetings that followed—a bargain ticket as far as I was concerned.

The third safe haven was the home of my Aunt Mary, my mother's second eldest sister. A woman of extraordinary human grace, Aunt Mary was married to a man who made a good living as a plasterer in Salt Lake City. Their second-born child, my Cousin Julie Ann, arrived on earth a week before me and became my closest cousin companion.

After we moved to Logan, I was frequently taken to Salt Lake for overnight visits. I loved pretending to be a member of Aunt Mary's household, secretly grateful that Cousin Julie Ann always claimed to be too homesick to visit me in Logan. I was six at the time of my first solo sleepover, having been sent to sleep with Cousin Julie Ann because she was sick with the measles. As I understood my assignment I needed to catch the measles from Cousin Julie Ann before I was old enough to have babies. I didn't catch the measles then or ever; ditto for the mumps.

The first measles morning, as I entered the kitchen for breakfast, I smelled Aunt Mary's homemade bread in the toaster and watched as she slathered my piece with butter.

Handing me the toast with a cup of hot cocoa, Aunt Mary said, "Carol, it tastes best when you dunk your toast in the cocoa."

Dunking or anything like that was strictly forbidden at my house. Aunt Mary not only allowed me to dunk, she saw I needed permission and lovingly gave it.

Aunt Mary's house had three bedrooms and one bathroom. The bathroom, situated in the center of the house, had peach-colored fixtures and two doors—one opened onto the front hall, the other onto the rear hall leading to the bedrooms where the children slept. I was mystified by how spotless I always found this heavily used way station.

Years later, when I was struggling to keep my own bathrooms clean, I asked Aunt Mary about this phenomenon. She told me as soon as anyone used the bathroom, she just quietly walked through it and wiped it clean—about forty-two times a day!

Aunt Mary also confessed that during the first ten years of her marriage, she had been in an unacknowledged competition with her next door neighbor to see who could be the first one on Monday mornings to hang their wet laundry from the clothes lines in their adjoining backyards. When Aunt Mary's fifth child arrived, the competitive pressure became too great, so she simply switched her laundry day to Tuesday.

During my childhood, high heels were the only footwear I ever saw Aunt Mary wearing. This too was part of the Latter Day Saints female coda—you happily got up earlier than your neighbor to hang out your laundry while wearing high heels so you always appeared at your most attractive best.

Long before I asked about Aunt Mary's miraculous bathroom, she was my maternal archetype. I never heard her raise her voice. Mostly I remember her saying, "Bless you, darling" and my body melting as her words washed over me.

Aunt Mary strove for perfection in her homemaking, but not in a way that was noticeable or made her children uncomfortable. Her methods of control and management were in stark contrast to those of Afton Evans. Only the marital bed was off-limits to the children. Like Afton's, its white chenille bedspread was stretched tight and smooth across the mattress. A needlepoint pillow of the Salt Lake Temple was always in the center of the pillow roll.

On sleepovers I shared Cousin Julie Ann's twin bed. I was still wetting the bed when I was eight, but whenever I was at Aunt Mary's house I somehow managed to wake up in time to use the peach-colored toilet. One night I found the back hall door to the bathroom was locked. While waiting to hear someone exit, I went back to bed and fell asleep. Cousin Julie Ann's bed was wet in the morning. When Aunt Mary asked me about it, I was flooded with shame and denied responsibility.

In an act of great compassion, Aunt Mary said, "You know, Carol, your Heavenly Father and I love you. You are always safe and welcome here no matter what. Bless you, darling!"

As a child in Utah, I never found homes where I felt as safe as I did in Afton's or Aunt Mary's, but I saw love in the eyes of dogs and cats and their soft warm bodies partially met my need for physical contact. An imaginary guardian angel carefully listened whenever I wrote in my diary, just as I hungrily absorbed the stories of others in books and motion pictures. Music often drew me out of self-doubt, allowing me to soar. I loved playing outside, away from home. I slowly developed an innate sense of what provided a safe distance between my mother and me.

-9-

The Great Fried Egg War of 1949

For my sixth birthday I was allowed to adopt two female kittens from the same litter of barn cats. I named the two feline sisters Blackie and Snowball. My father was initially against the acquisition. I no longer remember my mother's position, but she must have agreed because she was then all-powerful. I lobbied hard using standard childhood methods—manifested delight in anticipation of parental permission, mixed with solemn, unrealistic promises to take full responsibility for their care. The two kittens were all I wanted for my birthday. Plus they were free.

Dressed-up Blackie and the author, Logan, Utah, 1948.

The addition of two kittens to our household was possible because the postwar boom was in full swing and housing stock was no longer as tight as when we first arrived in Logan. After a year-long search, my parents located a two bedroom rental house only four blocks away from Marlene. Built by a young farmer, whose wife had died mid-construction while giving birth to their first child, the house seemed to reflect the builder's sense of abandonment. It sat on an unusually high cement foundation, as if it too was waiting for a lift-off to heaven. Because of the disproportionate distance from door to earth, egress was a challenge. It required careful maneuvering up steep, poorly formed cement steps to small unprotected thresholds at both the front and back doors. There were no roof overhangs to buffer weather impacts at these precarious entry points.

The house was on N. 500 West then near the southern city limits of Logan, one alfalfa field away from the railroad tracks. An adjoining lot on the east side of the house was the site of a cinder block warehouse filled with building materials and mechanical castoffs. On the west side was another field of alfalfa. Directly across the street to the north were the homes, sheds, and barn of Linda Olsen's family.

I was sitting on the front steps when Linda, who, like Marlene, was two years older than me, boldly crossed the street, brown hair pulled back by blue barrettes, wearing jeans and a torn short-sleeved plaid shirt. Hands on her hips, she asked me if I wanted to play hopscotch. Maybe because there was no sidewalk on her side of the block, Linda acted as if she was dependent on me inviting her onto mine, even though she had the chalk and only she knew how to draw the grid. The first thing I learned about Linda was she knew how to get anybody to do anything she wanted.

Linda's family lived in a kind of mini-village. Her parents, she, and her baby brother lived in one structure. Next to it was the larger house her grandparents had built almost fifty years earlier. Twelve children had been born in that house after her grandparents had been given the deed to the property by her great-grandparents for one dollar. Several of her father's siblings were still living there. There was a barn, a blacksmith shop, and sheds with mysterious contents. Betwixt and between was an aged horse, several cows, a coop full of chickens, four hunting dogs, and a colony of cats. Some members of Linda's family were Jack Mormons. Linda's grandfather was well known for his consumption of liquor and some of his children had fallen into non-Mormon habits like sipping coffee and smoking Camels, which made me feel more comfortable with the spiritual and genealogical imperfections of my own family.

Blackie and Snowball were from a litter of barn cats belonging to Linda's grandmother, Alice. Blackie was all black with short hair and classic yellow eyes; in contrast, Snowball had long white hair the texture of soft angora and unusual pale blue eyes. The black and white sisters seemed very happy to become my feline friends. They purred as they snuggled with me, licking my hands and legs. They slept together in a wooden crate Daddy wedged into a rough niche beneath the back door stairs. I made it soft and cozy with one of the surplus Army blankets Daddy always kept in the trunk of his Chrysler.

The basement of our new rental house, like many others in Logan, had been converted into a post-war apartment and rented to a young couple before we moved in. Stan Young, a student of animal husbandry at Utah State University, was hardly ever home, but I spent many comforting hours with his wife, Elaine. She was almost six feet tall, big-boned, with thin, wispy light brown hair. Elaine's most noticeable feature was a large mole shaped like a brand new pencil eraser in the center of her forehead, so prominent it gave her unusual authority for a woman. Totally ignorant of reproductive profiles, and fully focused on her forehead, I failed to notice that Elaine was pregnant.

Elaine was raised in Idaho in a large Mormon family and had very strong fundamentalist beliefs about the relationships between humans and animals. One of them was that cats should only be fed out of tuna fish cans, so I served all food and water to Blackie and Snowball in such cans.

Before we had even finished unpacking, my mother started talking about building a "Dream House." We soon began driving around Logan on weekends looking for "just the right place." My mother was not interested in the periphery of Logan, where fields of alfalfa were being turned into new building lots and ready-made tract homes; she was adamant it had to be a place where she could know in advance exactly what would be proximate to the boundaries of her Dream House. It took almost a year, but she finally found a spot* she liked on the opposite side of Logan, a lot that had once been the prized garden of a professor of range management at Utah State University. Since it was in a well-

---

*The garden property was located at 620 North 400 East in Logan.

established neighborhood, the only development surprises would be my mother's. My parents bought the garden.

A Dream House was the first thing I remember my mother wanting in ways I could readily observe. Up until then I believed she had wounds of dissatisfaction hidden beneath her stylish dresses, but I had no idea how they might be treated except by the acquisition of sable coats and canned crabmeat.

My mother went to work at Sears Roebuck so money could be saved to actually build her Dream House. In a stroke of personnel management genius, Sears Roebuck put my mother to work behind the returns counter at the back of the store. Right away my mother was recognized as someone fully familiar with dissatisfaction.

My parents worked semi-staggered shifts. Daddy was the first to leave in the morning; Mom was the last one home in the evening. This gave me several hours after school to spend with Elaine, who became my most important life-science teacher.

Elaine taught me why tomcats try to kill boy kittens, even if they are the daddy of those kittens. Elaine also told me where human babies come from, while showing me a menstrual pad, and providing a general description of their purpose. This was extremely interesting information! Her afternoon tutorials were far more useful than any of the lessons taught at Ellis School—plus Elaine was born to teach.

The snowfall in Cache Valley during the winter of 1948 and 1949 is still among the heaviest on record. There were six-foot-high piles of snow at the edge of our sidewalk; hay drops were organized for starving cattle, sheep, deer, and elk trapped in the deep snow all across the state.

On the first Sunday of December 1948, as Marlene and I were walking home from the Ward, she told me the truth about Santa Claus. The only thing on my Christmas wish list was a Toni Doll. Marlene already had one. I had enviously watched her rolling up the long hair of her Toni Doll in miniature curlers, after which she deliberately dabbed the rollers with a fake solution to give her doll a Toni home permanent. While debunking the myth of Santa Claus, Marlene was kind enough to assure me that she was certain my parents would get me my own Toni Doll. Not only that, they might have already done so.

Marlene Evans and the author, Logan, Utah, December 1948

While Mom was getting gasoline and Daddy was chipping ice off the sidewalk, Marlene and I stacked one of our kitchen chairs on top of a living room end table. As I held the chair steady, Marlene managed to lift the cover board on the entrance to the attic. She reported that she could see a big box laying on the rafters above my bedroom. Through the cellophane window in the lid of the box, Marlene saw a dark-haired Toni Doll in a blue dress with white lace that would be mine on Christmas morning.

December 1948 seemed endless. It was the combination of the cold, the steady snow, the recent death of Santa Claus, the preoccupying presence of the Toni Doll laying longingly above my head, and the looks of Snowball and Blackie as they sat on the narrow brick ledge outside of the kitchen window, pleading with their eyes and meows to join our warmth. It was so cold they were desperate to get inside the house, but no animals of any kind were allowed to come inside my mother's house. This combination of circumstances made me look forward to a new year.

As it turned out, 1949 was a very important benchmark in my life. On February 19, 1949, as I dressed for Primary School, instead of wearing my heavy rubber boots, I slipped on my patent leather Sunday-only Mary Janes because the winter sun had finally melted the snow from the roads and sidewalks. Pushing open the front screen door, I kept my eyes focused on Linda's house because we usually walked to Ward together. Eyes up, I accidentally stepped on Blackie's first litter. She had given birth on the front door mat and abandoned her eight kittens. All but one was still inside their birth sacks and dead by the time I stepped on them. At first I wasn't sure what the kittens were, but I knew they represented something awful, something I instinctively didn't want to see, yet I couldn't stop looking.

I mistook the one sack-free kitten for a field mouse. Daddy came to my side and tried to explain what must have happened, as he lifted me back inside and gently removed my Mary Janes. I ran to the bathroom and vomited. Then I took off my Sunday dress, lay down on my damp bed in my slip, and fell asleep.

A few days later Elaine discovered that Snowball had also given birth. She had two kittens hidden in a pile of lumber stacked on the west side of the warehouse. On Monday morning, while Snowball was having her breakfast by the backstairs, Daddy used his flashlight to show me that one of Snowball's kittens was a calico and the other black with white paws.

Eventually Elaine took matters into her own hands, moving Snowball and her kittens to a rag-filled box by the door to her apartment at the bottom of the basement stairs. My mother continued to be adamant that all animals belonged outside, but Elaine worried that the neighborhood tomcats might go after the black one with white paws because he was a boy kitten. As my ultimate authority in feline matters, I felt very relieved Elaine found an indoor solution in her territory.

But on that freezing morning, I went back inside and sat down at the kitchen table, waiting for my mother to fry me an egg for breakfast. As I waited, Blackie jumped up on the brick window ledge by the kitchen window, pleading to come inside. Looking into Blackie's yellow eyes, I remembered her abandoned litter and lost my appetite.

When my mother came to the table with the fried egg, I said, "I'm sorry, but I'm not hungry anymore."

My loss of appetite started The Great Fried Egg War of 1949.

Enraged, my mother said, "It is not acceptable to ask for something and then change your mind! It is wasteful and will not be tolerated. You may not eat anything else until you have eaten the egg you requested!"

As I left for school, my mother handed me an unusually light lunch sack. At lunch time, I found only the cold fried egg wrapped in waxed paper. That was the first moment I understood the nature of war between human beings.

I have no explanation as to why I failed to throw the egg away or why I didn't lie—telling my mother I had eaten it. Maybe I was too afraid, too stubborn, or too enraged, or all three. I rewrapped the egg, brought it home from school, and placed the sack on the kitchen table to await my mother's return from Sears Roebuck.

It was only my mother, me, and the fried egg for dinner. Daddy didn't come home because he had left for Southern Utah on a five-day DDT and Dieldrin spraying assignment. Sitting across from my mother at the kitchen table, with the familiar fried egg on my plate, I watched her eat a whole can of Del Monte "Sunday-only" fruit cocktail, but did not touch the fried egg.

The life span of a fried egg is extremely short—maybe only five or ten minutes. The fried egg was definitely dead and looking worse at every meal. I faced it down at breakfast the next morning before taking it to school again. Now completely preoccupied by thoughts of food, I kept my stomach quiet by drinking my school-provided milk and finishing off several half empty cartons left behind by classmates. My head was heavy with a hunger headache by dinnertime. On my dinner plate the fried egg appeared to have become petrified. Relieved when my mother allowed me to go straight to bed, I drifted to sleep listening to the sound of her working a batch of bread dough.

The next day I unwrapped the fried egg during lunch period in front of Vicky Greaves, my former neighbor. The sight of it inspired her to give me half of her sandwich. With this handout, I lasted until Wednesday night when the fried egg was once again placed on my dinner plate. I was feeling more adequately armed because Elaine had given me two of her homemade oatmeal cookies and a glass of milk after school. My mother ate her meal before removing the batch of bread dough from the refrigerator. She began kneading the dough into rolls for her cousin's family because her cousin was in the hospital.

It was the scent of her homemade rolls baking that forced me to disarm. I went to the table, closed my eyes, and ate the fried egg in one gulp, shutting down all bodily sensations.

Seeing that I had fully surrendered, my mother's mouth formed a pseudo smile, but she said nothing. I requested permission to immediately consume several handfuls of raw bread dough. When the first batch of rolls came out of the oven—I gobbled down three of them.

A few hours after The Great Fried Egg War of 1949 came to an end, a powerful chemical reaction began inside my body. It was similar to something I would discover a year later also existed in Yellowstone National Park. It was as if my body had become a pulsating mud geyser belching sulfurous smoke. I was sick through the night. For days afterwards, deep rumbling emissions of malodorous gas continued to rise from the depths of my body, pouring out from my mouth and nostrils in a visible vapor.

Over the next decade the sulfurous mud geyser symptoms reoccurred about once a month, always after I ate something prepared by my mother. This phenomenon came to be known in my family as "Carol's Icky Burps." I learned to run for the nearest source of fresh air to spare those around me from the silent, but not scentless, memories of the day I finally surrendered and brought an end to The Great Fried Egg War of 1949.

I did not eat another egg of any kind until I was twenty and on the opposite side of the Atlantic Ocean from my mother.

-10-

Baptized a Mormon in Sinful Apostasy

On June 24, 1950, I left home alone and walked ten blocks north to the Logan Temple to be baptized a member of the Church of Jesus Christ of Latter Day Saints.

After several years of preparation in Primary School, I was well-versed in the rules of Mormon baptism. Six months earlier, I had been chosen to play the Virgin Mary in my Ward's Christmas program. My selection for this role now seems suspiciously strange. But more importantly, I had reached my eighth birthday, an age at which Mormons believe a child is old enough to know the difference between right and wrong.

The author in head scarf, (third row, to left of podium) played the Virgin Mary in Primary School Christmas Program, December 1949

I went to the Logan Temple alone because only "Mormons who are in good standing" who have a "Bishop's Temple Recommend" may enter. Those two requirements eternally excluded my mother and father. Because our minds operated in three separate realities, neither of my parents expressed concern or disapproval.

The doorman told me to wait in the entry area. An older woman soon came to escort me to a dressing room, where she gruffly instructed me to take off my street clothes, while laying out a modest cotton garment resembling an oversized flour sack. After another period of waiting, she returned and led me to the baptismal font was where two Mormon elders lowered me into the water under the watchful eyes of golden oxen.

Baptismal font surrounded by golden oxen in Logan Temple

Regretfully, I was already in a serious state of disbelief about the cleansing power of baptism. I knew that I had already fallen from grace. While my childhood moral compass was not yet fully magnetized, I did not believe baptism would cleanse me of my sins as promised because I had known what I was doing was wrong even though I was underage. My mental state under water foreshadowed my future apostasy.

My fall from grace began on Mother's Day, a month before my eighth birthday. I hadn't remembered it was Mother's Day until the Bishop mentioned it in his opening prayer. Husbands and children stood up to testify about the gratitude they felt for their wives and mothers. In the midst of this powerful cultural chorus, I suddenly wanted something to give my mother. After Primary School and Testimony Meeting ended, I stood in line with other members of our Ward as we began moving slowly through the foyer to the front doors. Members of the Relief Society were distributing small potted pink geraniums in green and white striped foiled paper wrappers to each mother. This was the perfect solution to my gift dilemma! I stopped in front of Verba Greaves, Vicky's mother, who was handing out plants on my side of the line.

Looking Mrs. Greaves in the eye, I politely asked, "May I please have a geranium to take home to my mother?"

From the other side of the table Vicky's mother announced in a loud operatic voice, "No, Carol Lieberman, you may not have a geranium because your mother does not deserve one!"

Verba Greave's condemnation punched me in the stomach. Cheeks burning, my eyes filled with tears, I stumbled down the front steps of the Ward and out onto the sidewalk. As soon as my breath came back, I began running.

I was only a block away before I realized Vicky must have told her mother about the fried egg. My hunger and preoccupation during The Great Fried Egg War of 1949 had weakened my normal alertness, my way of keeping a no-trespass space between my parents and the surrounding Mormon world. The desiccated fried egg was consistent with what Verba Greaves believed was likely to happen if a beautiful Mormon woman married a Gentile and stopped being an active member of the Mormon Church.

Out of breath, I slowed my pace and began picking flowers growing out of various fences, defiantly breaking the eighth commandment not to steal. Grabbing and severing the stems of tulips and poppies, all I could see was the look of disdain on Verba Greave's face. As I gathered my bouquet of revenge, I realized I was deliberately provoking my Heavenly Father.

Looking upward, I shouted, "Heavenly Father, you are no help at all! You are meaner than my mother!"

My outburst broke the third commandment not to take the name of God in vain.

Inside our house, I heard the sound of the toilet flushing and found a note from Daddy on the kitchen table saying he had gone to check the temperature in the specimen room. Before Mom came out of the bathroom, I dropped the stolen flowers on the table and banged out the back door. Trying to distract myself from humiliation, I began walking on the rails of the railroad track behind our house in my Mary Jane patent leather shoes breaking the fifth commandment to honor my parents by disobeying one of my mother's strict Sunday-only shoe rules.

The day after Mother's Day, Mrs. Hill, my third grade teacher at Ellis School, decided our class was being disrespectful. She ordered an hour of absolute silence and said anyone who spoke would be held after school.

I hadn't been able to trust Mrs. Hill since the second week of school when she discovered me hiding in the bathroom after school and forced me tell her why. Royce Larson, the biggest boy in our class, had threatened to take me behind the garage of Widow Johnson, who was both blind and deaf, where he said he was going to tear off all my clothes. I was terrified! After I begged Mrs. Hill to call my father to come pick me up, she instead told the principal of Ellis School, who drove me home.

The next day, all the girls and all the boys in my class were marched into the music room. Girls were lined up facing the boys who stood opposite, while the principal lectured us on moral behavior and the importance of physical modesty. Royce leered at me through the entire lecture.

So, on the day after Mother's Day, during the hour we were paying for the sin of being disrespectful with restless silence, I went up to the big crayon box at the front of the room. Vicky Greaves followed me.

"What color are you looking for?" Vicky whispered.

"Black," I answered.

In a loud, sharp voice, Mrs. Hill said, "Carol, you will stay after school for one hour!"

I was humiliated and terrified. Even though Vicky spoke first, her violation was ignored. Now Royce knew I would be going home late and alone. Urine trickled down my leg. Seeing the puddle at my feet, Vicky made certain that by the end of the day everyone in class knew that I had peed on the classroom floor. I don't remember how I got home that day, but I know that I never spoke to Vicky again.

From the beginning of our friendship, Linda Olsen had insisted that I do my best to repeat everything Elaine Young told me. After watching Elaine nurse her new baby boy and Snowball nurse her kittens, Linda began to wonder whether milk would come from our nipples. She had "bud breasts" because she was, as my mother said, "old for ten." My own chest was flat. We decided to see if either one of us could taste milk from the nipples of the other after discovering the physical impossibility of reaching our own. Linda, having seen her grandmother's breasts, told me she was certain old women could reach their own nipples.

The day before my eighth birthday, we met in an old shed behind her grandmother's barn. The air tasted like the dust and linseed oil embedded in the wood. The sun seemed to be drilling through the warped openings, sending in revealing rays. A refrain from the hymn, "Jesus Wants Me for a Sunbeam" began pounding through my head. I was afraid that I was about to let Linda down, that milk would come out of her nipples, but not mine. Certain my Heavenly Father was watching, I knew He was not happy. My conscience was also shouting that an earthly person might discover us at any moment.

Since Linda was fearless, she went first. She was not certain, but thought that she could taste a little milk. With this possibility established, I moved my mouth to her bare chest. I struggled to put my lips on her left nipple, but my whole body was shaking. After a few feeble attempts, I moved away, admitting through my cotton-filled mouth that I tasted nothing. Even though Linda's skin tasted like cigarette smoke, that was something I knew not to mention.

Certain that there must be some milk in her nipples, Linda insisted I try again on the right nipple. The second time I lied, telling her that now I could taste the milk. That was the first of many lies I have since told about another human body. As soon as I knew what was desired, I told Linda what she wanted to hear, even though I had to break the ninth commandment not to bear false witness against my neighbor.

So, at the moment of my immersion in the baptismal font inside the Logan Temple, surrounded by the watchful eyes of golden oxen, I felt beyond cleansing. I had taken God's name in vain, stolen flowers, disrespected my parents by breaking my mother's rule about Sunday shoes, allowed myself to be filled with rage and revenge, done something in the shed I knew was a sin, and then lied to my friend about her body. Overflowing with feelings of shame, regret, and hopelessness, my hair still wet with Temple baptismal water, I wept my way home.

-11-

My Father's Feline Dream House

Using my new expertise from Elaine's extra-curricular life science program, I noticed both Blackie and Snowball were pregnant again. Snowball was now fully accustomed to her place by Elaine's basement door. It was not clear what Blackie would do.

A few days after Snowball's second litter was born, I could see from Blackie's profile that she had also given birth. Had Blackie hidden her kittens or were they laying somewhere in an awful mangled mess?

It was Elaine who saw Blackie enter the warehouse through a broken window. Blackie had moved one tiny step up the ladder of maternal instinct—she had hidden her litter in a barrel on a high shelf behind an old thresher machine. Blackie's kittens were safe, but they were starting life on a bed of nails.

When Daddy came home from work, I showed him Blackie's hiding place, and he immediately declared, "It is high time those two feline mothers have their own house!"

Friday night Daddy came home with lumber tied to the top of the Chrysler and worked past dark. He was back in construction mode just after dawn on Saturday. By the time I came home from Primary School on Sunday, my father's feline dream house was finished. It had three levels: an unfinished basement with a sand floor for the cats' personal needs; a main floor for eating; and a sleeping loft on the third level with a protective railing. The front wall of the house had a two-story opening covered with window screen, a feature that allowed us to see who was home and for cats to see who was calling.

We moved Snowball and her six kittens into my father's feline dream house Sunday night, putting her old Army blanket on the loft level for a familiar smell. The next morning we carefully moved Blackie's five offspring. This disruption was the perfect excuse for Blackie to abandon her kittens. She wandered away.

For the next two days Snowball struggled to nurse all eleven kittens by herself. Elaine mixed raw eggs with expensive real tuna fish for Snowball, worrying about the strain on Snowball's body. Elaine even gave me some of her breast

61

milk to put in my baby doll's bottle, coaching me in my efforts to feed Blackie's kittens, but they had no instinct for the smell of Elaine's breast milk. On the third day of Blackie's absence, the runt from her litter died. Elaine gave me an empty cornmeal box and an old washcloth for a coffin.

On the fourth day, Linda and I returned from a bike ride to witness a mothering miracle. Snowball was lying on top of Blackie with all ten kittens nursing in two rows stacked one on top of the other! With Elaine's expertise, we decided that because food was only available inside the house, Blackie had returned for a meal.

Snowball must have told Blackie in no uncertain terms, "Get up to the loft and lay down!"

Snowball persisted in this laying-on-motherhood-method of coaching Blackie for the next four weeks.

Disaster struck in the middle of the night when the kittens were almost six weeks old. Awakened by screams, I heard Daddy go out the back door. The sounds were terrifying. Stumbling down the back stairs into the darkness, I could hear Blackie and Snowball hissing and screeching, along with the frightened squeals of the kittens. Then came the sound of Snowball's body as she threw herself against the front screen. It was hard to see what was happening; only Snowball's beautiful white hair was visible in the darkness.

Daddy called, "Carol, get the flashlight!"

I ran back into the kitchen to retrieve it from the tool drawer and, as I handed it to Daddy, a large unknown tomcat brushed by my bare legs. Then there were only the sounds of mewing kittens.

Snowball was on the main floor licking one of the kittens. Blackie was on the back lawn cleaning her own wounds, which glistened in the beam of the flashlight. My stomach turned over because most of Blackie's left ear was missing. After we counted only four kittens in the feline dream house, Daddy began looking around the yard. The bodies of the six boy kittens were scattered from the feline dream house to the edge of the lawn. A large grey tomcat was still lurking at the boundary of the alfalfa field. Using rocks from the driveway, Daddy aimed and fired and the murderer disappeared into the dark field.

Elaine had been right: male cats killed baby boy kittens, and for all of Snowball's efforts, Blackie's mothering was no longer needed. The next day only Snowball was nursing. Blackie kept her distance, continuing to lick her own wounds in the niche by the back stairs.

Daddy and I formed the burial team. I marked each grave with Santa Monica sea shells sent to me by my great Aunt Genevieve, who had moved to California.

Blackie disappeared the following week. Elaine told me she was probably sick from infected wounds.

Rubbing my shoulders, she said, "I'm so sorry, Sweetie, but Blackie may have gone off to die."

It was not a time when you paid a veterinarian to treat a cat, at least not in my family. Two days later I discovered Blackie's body in the culvert where the irrigation ditch ran beneath the street. After Daddy dug another grave, Elaine gave me one of her flannel receiving blankets to wrap Blackie's body for burial. I still remember how her body felt stiff as a board. I was struggling to breathe through deep gasping sobs when Elaine reminded me that the last thing Blackie had done was to try to be a good mother.

"Blackie was a heroine, honey, remember that, okay? She gave her own life, trying to save her babies!" Elaine's voice cracked through her tears.

In the morning Elaine stood next to me holding her own baby boy as I made a gold stars out of yellow gravel rocks on top of Blackie's grave. When the four surviving female kittens were weaned, with Linda's help I found homes for them. Snowball lived on, but to me she seemed sad and lonely.

Elaine told me, "Animals carry our grief for us in strange ways," a phrase she repeated over and over until it became an essential commandment in my conversion to Elaine's mammalian doctrines.

At the end of August we moved into my mother's Dream House. It never occurred to me that Snowball wouldn't move with us, but she disappeared the night before the moving truck arrived. Desperate with grief, I rode my bike across town in search of Snowball every afternoon after school, while keeping one eye out for Royce. My searching slowed after the snow arrived, but Elaine and Linda kept looking for Snowball. Sadly, I never felt her beautiful soft coat or looked into her loving blue eyes again.

Elaine had told me many times that cats rarely bonded with people, only with places. But that didn't explain why Snowball hadn't been seen even once inside my father's feline dream house or at Elaine's door.

The feline dream house stood empty at the end of the drive for two more years. Then Elaine asked my father for permission to haul it to their new place near Caldwell where Stan was going to manage a dairy farm.

More than five decades later, when my father was near the end of his life, I asked what had led him to build the feline dream house. He took off his glasses to wipe away his tears, and with deep sobs heaving out of his skeletal chest, my father began his second-to-last pet confession.

"The day before we moved," he said, "your mother announced that she would not allow Snowball or the cat house to move with us. I argued with her, but she had made up her mind. Your mother put Snowball inside a packing box in the trunk and told me to drive out of town, past Hyde Park. When we got closer to Smithfield, I started to look for a barn since Snowball was a good mouser and I thought she would most likely find a source of water and winter warmth inside a solid barn." The front of his shirt was now damp with tears.

My father remembered there was a big irrigation ditch overgrown with tall grass that ran parallel to the road. He had twisted his ankle trying to jump across the ditch, then tore his new Arrow shirt on the barbed wire as he reached over the fence to put the box containing Snowball inside the cow pasture.

He continued with his confession, "I loosened the flaps of the box so Snowball could push her way out. After I turned the car around in the roadbed, I watched her through the rearview mirror. She came right out of the box." Still failing to suppress his sobs, my father said, "As I drove away, Snowball began running after us, but I lost sight of her in the cloud of dust."

As he always did, my father berated himself for his tears. "I don't know what is wrong with me—why I always start crying."

By then I knew that my father's heart had been broken in childhood and had never healed. I made him a cup of tea. Rubbing his shoulders, I told him I understood. As his emotions began to settle, my father told me more.

"I drove out to that spot many times over the next weeks on my lunch hour hoping to find Snowball. For months afterwards, I hadn't been able to sleep well. Your mother blamed my disturbed sleep on the strangeness of our new house."

One day he had approached the farmer who owned the barn, using his role as a USDA research scientist as a pretense for asking about the farmer's methods of rodent control.

"The farmer said he relied on his cats, so I asked him if he had a white one," my father explained. "He didn't, but when I saw how heartbroken you were, I thought if I could find Snowball, I could take her back to live with Elaine."

Our tea was cold by the time my father slowly shuffled down our front walkway to his Toyota Corolla. Then my own tears became a tender waterfall for the man who once built a feline dream house for two mother cats and their kittens. All families engage in acts of betrayal and abandonment. The majority do so unconsciously, with mostly benign effects. But in some families, like my own, betrayal and abandonment cut a deep wide chasm, an unbridgeable distance between those on either side.

Snowball

-12-

The Bear Who Loved Cherry Chocolates

My parents stole a week away from building my mother's Dream House and their respective jobs in July 1950. We packed the Chrysler with army surplus camping gear and drove north to Yellowstone National Park. The Beck family had gone to Yellowstone almost every summer throughout my mother's childhood. After their engagement in 1940, my father joined her family on one of those trips. Then World War II and gasoline rationing intervened. Our trip in 1950 was my first and the first in a decade for my parents.

Even though we arrived after dark during peak season, there was no problem locating a campsite. Yellowstone was not yet teaming with tourists. We pitched our two Army surplus tents on the north side of Yellowstone Lake near Fishing Bridge. In the morning, my father helped me catch a lake trout and cook it over an open fire for lunch. It was the first fresh fish I had ever eaten. Until that moment, I had only tasted canned tuna. What a delicious difference!

After lunch we climbed into the Chrysler to look for animals. It was during the era when Yellowstone bears came to the roadside for human handouts. The whole country was so naive about nature there were no prohibitive feeding rules, nor any sense of shame. There were many other wild creatures in Yellowstone, but none were as interested in humans as the bears.

We started driving the Grand Loop Road, going north toward Canyon through the Hayden Valley, a spectacular place where the Yellowstone River flows leisurely through a mountain meadow dense with wild grasses. The Hayden Valley is still wildly rich in all of Yellowstone's creature attractions. That day we saw swans, bears, buffalo, moose, elk, Canada geese, ospreys, pelicans, and a coyote.

Afterwards we drove to Canyon Village. We all had a grilled cheese sandwich and a butterscotch malt at the lunch counter in the Hamilton Store. As I made "ill-mannered" straw sucking noises in search of the last drops of creamy sweetness, my parents decided it was too late to start the hike down the Grand

Canyon of the Yellowstone. Saving the trek for the next day, we climbed back into the Chrysler and continued on the Grand Loop Road to Old Faithful.

We sat on long logs with other tourists waiting for the great geyser to put on its hourly show. When the magnificent spout retreated for another rest, we made our way to the Old Faithful Lodge for ice cream cones. Not a fan of ice cream, my mother went into the gift shop and bought a box of cherry chocolates.

Back at our campsite, my father proposed we get up early the next morning to see more wildlife before hiking down the Grand Canyon. His idea prompted my mother to suggest we eat breakfast at the Hamilton Store rather than trying to cook and clean up so early.

At dawn the next morning, we once again started driving north on the Grand Loop Road alongside the Yellowstone River. We soon came upon a very large mother bear with twin cubs sitting on the opposite side of the road. The rising sun was shining on the trio and the mother bear's coat was almost as auburn as my mother's hair. It was a spectacular sight, made especially exciting by our solitary status. My father pulled the Chrysler to the side of the road directly across from the bears and turned off the motor so we could hear any sounds the trio might make.

Watching the twin cubs attempting to suckle gave me a queasy feeling in my stomach—both Blackie and Linda Olsen popped into my internal viewfinder. The mother bear kept roughly batting her cubs away while keeping a careful watch on our car.

My mother slowly rolled down her window. In 1939, when my father's Chrysler was manufactured, all cars had crank window handles, and ours had become stiff and hard to turn after more than a decade of use. My mother unwrapped one of her cherry chocolates, squeezed the upper half of her slim body out the window, and tossed the cherry chocolate over the car roof into the far lane of the roadbed.

Instantly the mother bear came down the embankment to retrieve the cherry chocolate. She had hardly tasted the offering when she charged across the road to my mother's side of the Chrysler, demonstrating she was both an experienced beggar and careful observer of the roadside scene. In seconds the huge mother bear was standing on her hind legs, looking directly at my mother through the open passenger-side window. Her nipples looked considerably bigger than any Linda Olsen or I had ever imagined.

My father yelled, "Margaret, roll up your window! Quick! Hurry! Hurry!"

Despite my mother's frantic cranking, there was still an opening of a few inches at the top of the window. The mother bear moved closer, placing her front paws on the top edge of the window glass, her two-inch-claws protruding into the car, like a row of ivory swords.

My mother began screaming at the top of her lungs, "Drive away! Drive away! Help me! Help me!" frantically leaning away from the bear, toward my father.

Yellowstone Bear Begging at Car Window

The bear yanked my mother's window down in a single swift effortless motion. But instead of turning on the motor, my father grabbed the box of cherry chocolates from her lap and threw the whole box out the open window, elbowing my mother in the neck as he flung the bear bait away.

The box grazed the side of the mother bear's head. She turned, dropped her front paws to the ground, and quickly consumed the remaining cherry chocolates, still in their individual red foil wrappers. My father started the engine and the Chrysler's wheels spun gravel as we roared away.

My heart was pounding! I was filled with awe at the strength of the mother bear—at the power the bear had over my mother! From behind the Chrysler's tall grey mohair front seat, I could only see the top of my mother's auburn hair, but I could hear her choked sobbing, mixed with gasps for air, and smell her fear. The scene felt uncomfortably close.

As my father continued driving north on the Grand Loop Road, the crisp morning air, barely warmed by the rising sun, moved through the broken window, over my mother, into my nostrils. The air inside the Chrysler was filled with a mixture of memorable scents—ursine fur, the scent of white firs, my mother's fear, mixed with her Coty's Emeraude perfume. A battle began raging inside my mind. Now I would describe it as awe overpowering fear, elation shouting down shame. I felt strangely excited by my mother's close encounter.

Grand Canyon of Yellowstone, 1950

My mother did not hike down the Grand Canyon of the Yellowstone. When my father and I returned from the deep gorge vistas, she adamantly announced she was done with Army surplus tents. So we passed through the Hayden Valley without pausing, packed up our campsite, and drove to Old Faithful, where my father rented a small cabin with four bunk beds. It also had a much needed indulgence—a small shower stall.

On our last day in the park, we drove north to Mammoth Hot Springs. Daddy and I spent the day walking around "Carol's Icky-Burp" smelling geyser system, while my mother sat in the Mammoth Hot Springs Hotel reading through a stack of ancient *National Geographic* magazines.

My mother insisted on driving, both to Mammoth Hot Springs and on the return to our Old Faithful cabin. She always drove fast, but on that day she never even slowed where bears were begging.

-13-

My Mother's Dream House

Of my seven childhood homes, I remember my mother's Dream House most vividly. It was custom built on a half-acre lot that had once been the garden of a Utah State University professor. Twenty years earlier, the professor had planted a diverse collection of fruit trees, berries, roses, exotic irises, and bulbs of all types. The lot was one of the few times my father paid above-market price. He recognized the value of the property's botanical attributes.

When construction began, the garden was brutally injured, even though my parents sited the Dream House closer to the street than the homes of our neighbors trying to protect as many of the plants and trees as possible.

I have no idea why my mother wanted a white brick house. I don't recall any other white brick houses in Logan at that time because there were no light bricks manufactured in the area. The contractor was adamant my mother would have to settle for the ubiquitous dark burgundy bricks. Although I was only nine years old, even I knew that wouldn't suffice.

I was with my father the day he happened upon a large number of used yellow bricks in Smithfield. He had to hand-scrape the mortar off the bricks because the contractor refused. I sat on the ground at the foot of what seemed to be a mountain of yellow as my father chinked away at the old mortar after work until it was too dark to see. I tried to be helpful after each brick was free of mortar by carefully stacking them near the frame of the emerging house. After the brick layer finished his work, my father painted the bricks white.

Part of me wanted to ask him if he was tackling the insurmountable mountain of yellow bricks because he loved my mother or feared her. But such a question would have violated our unspoken boundary. However, I believed my father had undertaken the chore because he had no choice; it seemed to me that my mother got whatever she wanted. Nonetheless, an emerging sense of romance, garnered mostly from Doris Day and Rock Hudson movies, made me wish that true love was the force behind my father's chisel.

Another of my mother's requirements was floor-to-ceiling windows on the east wall of the living room. She got the tall east windows and the white bricks, but other compromises were necessary. My mother wanted exterior shutters on the front windows. The retail cost was too high, so my father made them from scratch, spending months working at night in bad light in our garage trying to create the louvers using inadequate tools. When he finished, my mother painted his carefully sanded shutters deep burgundy which she said "dressed up" the white brick.

My Mother's Dream House, 620 North 400 East, Logan, Utah 1956

My father was born with the heart of a gardener. As the construction neared completion, he began restoring the garden, maintaining it in near flawless condition.

What I most remember about my mother's relationship to the garden was her daily breakfast-timed announcements, "Last night, I saw the ghost wandering through the garden." A repetitive report uttered even when the garden was buried in snow.[8]

I had anticipated my mother's Dream House would feel perfect, but our furniture suddenly seemed old and sparse in the way Ben's birthday couch had made the Evans' living room seem shabby. Of course, there was no money left for new furniture or other interior features. My parents retrieved some used grey carpeting being torn out of a fraternity house, self-installing the recycled carpet on the living room floor.

That fall I would come home from school to find my mother struggling to make pinch-pleat drapes using the directions from an issue of *Better Homes and Gardens*. The only fabric within the household budget was plain white sheets from Sears Roebuck, where my mother still had an employee discount for working in returns after Christmas when the load of dissatisfactions peaked.

On the rare occasions when a friend came inside our house, I would say we were waiting for our new furniture to arrive. I regret having engaged in such decorating deceit, but I now know now that strong design preferences are one of my maternal inheritances.

Unable to afford to finish the upstairs rooms, my mother's Dream House had a central staircase that went nowhere the entire time we lived there. Nor could my parents afford cement for the driveway; my mother had to settle for gravel. A fireplace was boxed in, but remained cold.

Our neighbors to the west were Thomas and Donna Nielson; who at that time had two little girls, Tana and Teri. It wasn't long before I had a regular babysitting job on Saturday nights. Donna insisted I was to sit in their room until the two girls fell asleep, just as she always did. Every minute I spent in that house seemed like ten. As soon as Donna opened the side porch door to let me inside, I felt desperate to leave. Yet, whenever Donna came to our front door to ask me to babysit, I felt powerless to refuse.

Although babysitting was my first source of independent income, it filled me with a strange sense of servitude. I felt frozen, as if moving a single inch in the wrong direction would result in disaster, which was entirely possible since another of Donna's rules was that I was not allowed in the living room—thus making it irresistible. As soon as the girls were asleep, the forbidden room called out to me. Dropping to my knees in the center hall, I would crawl into the living room to admire Donna's extensive collection of porcelain ballerina figurines. Some had skirts made of real tulle which had been dipped in glaze and fired. On either end of the Nielson's rose-colored couch were tables with twin lamps, the bases of which were shaped like toe shoes. Reaching a good point of observation, I would turn over on my back and dream of dancing in a pink tulle skirt like some graceful princess I already knew I would never become.

In the 1950s, inhabitants of Logan rarely closed the drapes on their front picture windows. Walking down the street at night was like going to a free peep show. The living rooms were well lit, but it was rare to see anyone living in them. Strangely, these empty tableaus were indicators of normalcy. Each had a stage-like potential, as if the play might begin at any moment, although it was definitely unseemly for the strolling audience to stand around waiting for the entrance of the actors.

Two things were different about our house. The drapes on our front picture window were always closed and those passing knew its occupants were Gentiles to each other.

We lived in my mother's Dream House for six years. During the second year my mother obtained a part-time secretarial job working for Professor William F. Sigler[9] at Utah State University. Her wages meant that the living room couch could finally be re-upholstered. When the couch returned from the furniture hospital, it had been "modernized." Something about it reminded me of one of the World War II veterans who had experienced amputation. My mother had chosen a dark green nylon pile fabric that I found stiff and unforgiving. I silently prayed that the poisonous snakes had also been removed.

Whenever I was home alone, I liked to listen to the radio while writing in my diary. I wrote in my diary as if my words meant something—as if my imaginary friend or a guardian angel or someone like Aunt Mary was listening to me, understanding all that I felt, and, as imperfect as I was, loving me, encouraging me to live on.

Now in my tenth year, I knew my mother was both dangerous and deeply disturbed. Believing her condition was somehow my fault, it was my duty to fix her. My constant secret worry was that I had no idea how to make the repair.

-14-

Branded for Life: My Atomic Tattoo

The move to my mother's Dream House meant I would start fourth grade at the Adams School, located in the adjoining block. But there was a hitch. Logan was bursting at the seams in the early 1950s because World War II veterans had arrived to study at Utah State University on the GI Bill.[10] They brought their wives and war babies—so many that a village of Quonset huts had been created on the north side of the campus to shelter them.

To ease crowding the school board decided to bus all the fourth grade students from every elementary school to Woodruff Hall, an older vacant vocational school building. Every morning I walked across the street to the Adams School and boarded a bus that took me across town to Woodruff Hall, where I had daily encounters with my old nemeses, Vicky Greaves and Royce Larson. Vicky and I ignored each other; Royce had moved on to more interesting bodies.

I didn't mind the bus ride and I loved my beautiful new teacher, Mrs. Barbara M. Hales. Best of all, on the second day I made a new friend, Sherry Watkins, who lived just three blocks from the Dream House. We began going to Ward together, as well as sitting next to each other on the school bus.

In addition to busing, my fourth grade school year was permanently marked by Utah's religious-like commitment to civil defense. Mormons have believed in disaster preparedness ever since those first lean and locust-plagued years in the valley of the Great Salt Lake. That was when Brigham Young received a revelation from God that we should stockpile! Mormon doctrine calls for every family to have a one-year emergency supply of food on hand.

So when the federal government started pushing civil defense readiness as part of the Cold War, the propaganda resonated through Utah like a voice from heaven. Inside Woodruff Hall, like most American school children, I learned how to hide beneath my desk with my arms over my head at the first sound of the fire bell.

One day I carried home a pamphlet with directions for constructing an underground fallout shelter out of cinder blocks. My father did not dig a big hole in our backyard garden. Having been sent by the USDA to sample bees and other insect populations in the Nevada atomic-testing zone, Frank V. Lieberman had a very clear picture of the limits of underground shelters made of cinder blocks as a defense against an atomic bomb. Not only were there no bees to be counted, everything was incinerated and turned into invisible lethal radioactive dust then blowing over our home territory.

"Something terrible is going to happen at school" was the first rumor I heard. A week later, just before the final bell, we were told to eat a good breakfast in the morning because lunch might be late. Also, girls were told to dress for school "only in skirts and blouses, no dresses." It was long before girls were allowed to wear pants to school.

The next morning we arrived at school to find a row of green Army blankets had been strung on a rope across the middle of the lunch room forming a fragile partition. One class after another was called to go into the room in two gender-distinct lines in alphabetical order. It took quite a few minutes for us to get us organized by last name—only a few of us were proficient in surname spelling. Mrs. Hales seemed confused and nervous herself.

Once both lines were inside the cafeteria, the big double doors were closed and we were told to remove our clothing to our waists. This meant there was a line of shirtless boys on one side and a line of topless girls on the other. To compensate for our fear and budding body shame, we produced an unusual amount of nervous tittering and blanket punching. Of course, there were attempts by the boys to part and lift the partitions. Two men dressed in military uniforms and two women nurses acted as monitors, blowing whistles whenever there was an infraction.

At the front of each gender line were two freestanding hospital style screens. One-by-one, each of us was taken behind first one, where our finger was pricked and a smear of blood was put on a card which somehow determined our blood type. After what seemed both an eon and only a brief moment, the smell of fear mixed with the scent of burned flesh filled our nostrils. Those of us in the middle and the rear of the line instinctively moved back as our classmates in the front of the line began whimpering, screaming, or passing out on the floor. Some children became nauseated and the nurse monitors quickly had their hands full. I remember watching Mrs. Hales on her knees with stiff brown paper towels from the bathroom dispenser trying to wipe away a puddle of vomit.

In 1952, in Logan, Utah, as part of an effort led by Dr. Omar Budge to create a "Walking Blood Bank" the blood type of every school child was tattooed on their chest to ensure proper first aid in the event of a nuclear attack.[11] These inch-sized identifiers were permanently etched on our bodies just in front of the left armpit at the level of the nipple. Uniform placement was intended to assist rescue workers determine our blood transfusion needs in the midst of the anticipated chaos of nuclear war.

My "O+" tattoo has stupefied my doctors and mystified lovers because it has been essentially illegible from the time my breasts needed to hide behind a bra. My own "Atomic Tattoo" has an extra mark of distortion from the kind of fatigue that would strike even the best tattoo artist after a long day of working on resistant, terrified canvasses.

That day, when I returned home from school tattooed, my mother lifted my blouse to see the result.

She quickly jerked it down, saying, "That extra mark is how they will know they aren't supposed to save you!"

Where blood type would be tattooed [X].

Illustration from Chicago Tribune, August 1, 1950

-15-

Never Worry About Where Your Money Has Gone

Money dominated my parents' dinner table conversations for many years. I remember my father climbing up the Federal pay scale, "GS" (government service) step-in-grades, full grades, and cost-of-living adjustments until he reached the pinnacle. In 1952, when his official papers came designating him a "GS12" as Station Chief, his annual salary was five thousand forty dollars. The repetitive dollar dialogue between my parents had taught me not to ask for anything. The price was too high.

Fortunately, I soon had my own money from babysitting the Nielson girls. I spent it wildly, mostly on movie tickets and candy. It was not until I started fifth grade at the Adams School that I got into trouble about money. I discovered Jantzen sweaters in Roskelley's, a small women's boutique on Main Street. By skipping movies and candy for a month, I had enough money for my first Jantzen sweater. I chose a sea green pullover with a Peter Pan collar trimmed in white angora, a rickrack-like design detail that reminded me of Snowball.

As I approached my tenth birthday, my breasts were beginning to look like Linda Olsen's. I also felt like I was going to faint every time Scott Olsen (no relation to Linda) looked at me. Scott, a fifth grade classmate, lived in Hillcrest, a new housing development at the mouth of Logan Canyon. With hair so blond it was almost white and large pale blue eyes, something about Scott's albino appearance also reminded me of Snowball.

Soon after my Jantzen sweater purchase, I talked about Scott unabashedly at the dinner table without recognizing my enthusiasm was triggering an alarm. I missed the first clue completely. Physical contact had always been in short supply in my family, so I had developed a ritual of sitting on my father's lap after dinner, while my mother cleaned up and they finished their table talk, the preponderance of which involved money and property maintenance. The night I first spoke about Scott Olsen, my mother announced that I was too old to be sitting on my father's lap.

My father responded, "Oh, Margaret, cut it out! She is still a little girl!"

Coincidentally, during his lunch hour that day, my father had bought a new straw hat. My mother had the kind of squinty-eyed, tight-lipped look that told me she was not happy. Feeling fearful, I quickly got up from the safe haven of my father's lap.

Trying to break the tension, I asked, "Daddy, may I try on your new hat?"

The eight dollar price sticker was still inside so I pulled it off, handing it to him. In this awkward triangulated moment, a deal was struck. My father, who must have interpreted my mother's look in the same way, intervened with a proposal.

"If you keep this price sticker for exactly one year, I will give you the same amount of money, eight dollars. If you forget, I won't give you the money, but you can sit on my lap until your sixteenth birthday. But until a year from now, no more lap sitting!"

What a bizarre pact! But in our family, bizarre was normative. In miniature script on the sticker, my father wrote the date of redemption, March 27, 1953. Taking the price tag into my room, I stuck it inside my purple hanky box, a six-inch square box covered in quilted satin with a hinged lid containing five cotton hankies, a ninth birthday present from Aunt Mary, whose instruction on my birthday card was: "Never leave home without a hanky." The purple hanky box quickly became the place where I kept all my treasured mementoes, superficially hidden beneath Aunt Mary's hankies.

Two days later I walked across town for my regular piano lesson with Mrs. Meda Snow. It was the last Saturday of the month, the day my mother gave me ten dollars to pay for the next month's lessons in accordance with Mrs. Snow's strict payment policy. I do not know how I lost that ten dollar bill, but it wasn't in my pocket when I reached Mrs. Snow's house. Seeing how upset I was, she tried to assure me that I would probably find the missing bill on my way back home.

As I left, Mrs. Snow said, "You can pay me next week just this once."

Carefully retracing my steps home, I found nothing. The money was gone! This was a disaster! When I arrived after my slow search, not only had I lost the ten dollars, but I had also lost the courage to tell my mother. I postponed denouement in the hope that the ten dollar bill might magically reappear. After a few days, I put the loss in the back of my mind where I kept my less important worries.

It was not until I was walking to my lesson the following Saturday that I remembered. Mrs. Snow announced she would have to telephone my mother to inform her of the situation. I pleaded without success for another reprieve. When my lesson ended, I heard her dialing before I had even cleared the front steps. I didn't know how my mother would react, I just knew it wouldn't be good. But never, not even in my wildest imaginative fears, could I have predicted the plot hatched inside my mother's mind.

"I know what you have done with that money!" my mother announced as soon as I opened the front door of her Dream House. "We are going down to Wickel's Men's Store right now and get that money back! Get in the car and don't say a word!"

I was speechless, so that was no problem. My mother drove to Main Street and parked directly in front of Wickel's Men's Store, located on the same block as the Blue Bird Café. She slammed the driver's side door and stood on the sidewalk glaring at me with such coldness my buttocks felt frozen to the front passenger seat.

Wickel's Men's Store (far right) on Main Street, Logan, Utah 1952

After a few seconds, my mother screamed, "Get out of the car right now, young lady!"

Paralyzed by fear, I could not seem to make my body move. Finally, my mother stepped off the curb, opened the passenger door, and pulled me out of the Chrysler. Then painfully pinching my right bicep, she mother-marched me into Wickel's Men's Store.

I had never been inside before. It was small, with a centrally located counter for the cash register and packaging tasks. My mother's high heeled shoes sounded like wild, out-of-control hammers as they briskly struck the polished wooden floor. Still mother-marching me, she pushed me toward the counter where a gentleman of middle-age stood.

"Who are you?" she rudely demanded.

"Good afternoon, Madam. I'm Samuel R. Wickel, the owner. How may I help you?" he responded with crisp, naive formality.

I could see how he was trying to create a safe, role-oriented distance from this unknown customer right from the get-go.

"My daughter Carol has been putting money down here on a sweater for a young man who has taken her fancy," her tone demeaning. "She did this without my permission and I want you to refund the money now!"

The customer facing Mr. Wickel had clearly acquired some special skills in counter confrontation while handling returns for Sears Roebuck. The fierce look on my mother's face, embodying her deep distrust of the world, was enough to incinerate the good intentions of anyone in the immediate vicinity.

I wanted to die, right then, right there. But instead of taking me into his arms for eternity, my Heavenly Father gave me back my voice.

"Mom, I have never been inside this store before! I just lost the money somehow. Mom, really, I don't know how, I just lost it! I have never, ever even thought of buying a sweater for Scott. Honest, I promise! Please, Mom!"

My mother, acting as if she hadn't heard me, repeated her demand. Mr. Wickel firmly stated that he didn't recall my having ever been in his store, but he would certainly check his records. Asking the other clerk to attend to a customer who had just entered, Mr. Wickel began thumbing through a card file of store account records.

"What did you say your daughter's name was, Madam? I assume the account would be under her name or would it be under yours?" Mr. Wickel asked politely.

My mother snapped back with steely sharpness, "You know whose name it is under and don't try to pretend otherwise!"

Mr. Wickel took on color. I knew exactly how he was feeling and so I threw myself on top of the hand grenade my mother had laid on the counter between herself and Mr. Wickel. Like all acts of low heroism, I acted instinctively, without forethought, in order to save an innocent man, whose hands had begun to shake.

Leaning forward, I croaked in a tone of fake authority, "It would be under Carol Lieberman, Sir."

The instant the words were out of my mouth, I knew my goose was permanently cooked. My youthful pseudo-heroism had just fed my mother's paranoia a big cherry chocolate!

We were at the counter for what felt like twelve hours as Mr. Wickel dutifully checked his lay-away file and his daily receipts from the previous Saturday. My mother finally agreed to leave after Mr. Wickel promised to check with other clerks who might have assisted me and to call my mother with the results of his inquiries. By this time, she had told Mr. Wickel my confession of the account being in my name was all the proof she needed. She expected the money to be returned by the end of the week.

As she pushed open the front door of the store, she turned back with one final demand.

"If my daughter ever comes in here again, you are to call me immediately! Do you understand?" she said.

"Yes, Mrs. Lieberman, I do understand and I will certainly keep an eye out for her!" Mr. Wickel replied.

There was hardly need for surveillance. I never went inside Wickel's Men's Store again. In fact, I never even walked past it again.

Back in my bedroom, I wrote myself a note, "Keep your mouth shut!"

As I opened the purple hanky box to tape my self-admonition on the inside lid so I would see it daily, I noticed for the first time that the eight dollar price sticker from my father's straw hat had the name "Wickel's" in tiny print at the top. This was the kind of coincidence that made me believe there really was a Heavenly Father—a meticulous, obsessive, all-knowing-God, who maintained a large notebook containing a page with my name on it, where he made big black "X" marks for each of my mistakes; double big black "XX"s for the times I broke one of his commandments.

For the next twelve months, I faced my admonition and the Wickel's price sticker whenever I opened my purple hanky box. Because it was where I kept Scott Olsen's notes which I liked to reread before going to sleep, I had almost daily encounters with both reminders. The day before I was entitled to collect my eight dollar reward, I announced at dinner that my father better get ready to pay up. The next morning, I completely forgot about my claim ticket. Apparently my unconscious knew I wasn't ready to sacrifice the tiny bit of comfort I felt in my Daddy's lap for any amount of money. Sadly, the redemption period had been so long, it was now increasingly awkward for me to claim my booby prize. I never sat on my father's lap again.

That humiliating day in Wickel's Men's Store instilled me with a life-long hypersensitivity to salespeople. Fortunately, I have lived long enough to enjoy what for me turned out to be *optimal distance* in consumer transactions—the safe anonymity provided by catalogues and the Internet.

I have also put into practice the important lesson my mother inadvertently taught me that day in Wickel's Men's Store. Money comes and money goes, but I don't worry too much about why or where.

-16-

The Rewards and Risks of Giving

Even though Marlene Evans had revealed the identity of my secret Santa, I still believed in the spirit of Christmas. Our teacher proposed our fifth-grade class collect pennies and hold a bake sale so we could buy gifts for a family of five children whose father had died in a truck accident. Three of the children went to the first, second, and third grades at Adams School. The youngest were twins, too young for school.

We worked with the eager, erratic energy of ten-year-olds, raising twice the money we had hoped for from our badly baked brownies and Christmas cookies. We set out to shop for the children with a list provided by their newly widowed mother. At the end of our last expedition, the students keeping track of our shopping money announced we had acquired everything on the list, but we still had money to spare.

Worried about the widowed mother, I raised my hand and suggested we use the extra funds to buy a present for her. Our teacher said it was such a good idea, I could select the present. I chose a light blue embroidered cotton hanky at Sears Roebuck and a bottle of Coty's Emeraude perfume at City Drug Store.

Five of us were chosen to deliver the presents and sing to the children. We met on the front steps of the Adams School, giddily packing ourselves and the gifts into our teacher's Ford coupe. She drove north, all the way to the edge of Logan and parked in front of one of the smallest and oldest houses.

We clumsily made our way onto the porch—our arms full of packages wrapped in self-painted paper, drifting irregular chips of easel paint like uncontrolled dandruff. The door swung open before we even knocked and the mother silently ushered us into the dimly lit living room. We crowded together like a herd of frightened sheep, the silence broken only by the sound of our nervous coughs and the creaking floor boards. I can still see those five fatherless children on a couch near collapse. In the absence of a Christmas tree, we laid our presents at their feet.

Our teacher then led us in a too-high pitched rendition of "Silent Night" followed by "We Wish You a Merry Christmas". By the time we began singing the last stanza, my body was covered with goose bumps.

Afterwards I had my first experience with car intimacy when I was the last student to be dropped off in front of my mother's Dream House.

Taking my hand, my teacher said, "I love you, Carol. Merry Christmas!"

Bathed in her loving light, I made a vow. Whenever I feel love for another human, I do not hesitate to say so. Most recipients of my declarations are puzzled, some alarmed, a few clearly threatened. Nonetheless, I take the risk of rejection in honor of a teacher who saw I was starving for affection and tenderly fed me.

The next day there was a highly unusual Lieberman family outing. We climbed into the Chrysler and my father drove us to Ogden to shop for our own Christmas presents. I was to acquire a new winter coat; my mother was going to buy a new dress; I no longer remember what was on my father's list.

In an elegant clothing store, my mother tried on several dresses. She suddenly looked as glamorous as any movie star as she turned around in the bright lights of a three-way-mirror, surrounded by saleswomen and other customers, all nodding their approval. My mother seemed transformed; as if something miraculous had happened. This was my first glimpse of the power of a shopping high.

Unable to decide which dress to buy, my mother told the clerk to hold her three favorites while we went to have "a bite to eat." We ate open-faced roast beef sandwiches covered with thick gravy in a restaurant near Ogden's Union Station, while my mother decided on one of the dresses. She had something else she wanted to shop for alone, so Daddy and I were assigned to pick up the chosen dress.

Back inside the dress shop, Daddy got a mischievous look in his eye. He told the saleslady he wanted to buy all three of the dresses. He also asked her to put them into a single box so he could surprise my mother. That was the first moment I remember feeling my father wanted to transform my mother as much as me. We became co-conspirators in something wonderful. There was hope! It was Christmas! It was possible to find happiness, even joy, by giving presents to others who had broken hearts and minds. Giving was the best gift of all.

In an extremely rare instance of acceptance, Mom happily kept all three of her self-selected dresses.

The next Christmas, my whole life had changed. My mother's father, Pepa, as he was now called by his five children and fifteen grandchildren, was living with us after having suffered a major stroke. Nanie was terminally ill with cancer, emphysema, and congestive heart failure. Aunt Mary took Nanie home to nurse; my mother took Pepa. I now see this delegation of responsibility to my mother by her older siblings as an indication that they had little or no direct experience with their sister's paranoid schizophrenia.

Up until the moment Pepa moved into her Dream House, my mother had been in full crisis mode—the curtains were kept closed all the time, the phone, wrapped in a towel, was stuffed inside a kitchen drawer, and she rarely left the house. I thought her regression had been triggered by the Logan City Directory. The 1953 edition had been delivered just before school started. Like many city directories of that era, it listed every adult resident of Logan by both address and occupation, including either the given name of their spouse or their marital status

> Lieberman Frank V (Margt) collaborator USAC h620 N 4th
> East

Listing in Logan City Directory, 1953

My father's occupation in the 1953 edition of the Logan City Directory was listed as "collaborator." This was a reference to his role as a USDA research scientist collaborating with the faculty of Utah State Agricultural College (it was renamed Utah State University in 1957). My mother's interpretation was quite different. America was then at the height of the Cold War and McCarthyism.[12] She believed my father had been publicly labeled as a spy and a Communist. Further, she too had been implicated as an enemy of America by being listed in the directory as his wife. This episode is my first memory of what became my mother's life-long obsession with the Federal Bureau of Investigation (FBI).

Pepa's stroke had left him unable to speak and he was paralyzed on his left side. At age sixty-five, he was only able to move with great difficulty and needed a wheelchair. Eating was taxing, but the greatest challenge was his inability to communicate. Whenever Pepa tried to speak, he burst into tears. Fortunately, my mother seemed to become more functional nursing her father. The curtains were once again opened to sunlight; the phone was back on the kitchen counter, and after feeding him lunch, my mother usually pushed Pepa around the block.

In preparation for Pepa's arrival, I moved to a room in our unfinished basement, a location I learned to like because it kept me at a greater distance from my mother. Our second-hand upright piano was in the same room so I found myself practicing more often. Since our only bathroom was located between my parents' bedroom and what was now Pepa's, my father suggested I use a chamber pot in the middle of the night, rather than trying to make my way up the steep staircase in the dark. I came to appreciate that solution after our bathroom became crowded with Pepa's equipment and supplies.

When the Beck sisters closed up the apartment of their parents, I inherited the bed and linens of my grandparents, as well as their radio. My father ran a wire out the basement window to improve reception. I fell asleep each night between sheets scented by the skin cells of my beloved Nanie and the always remote Pepa, while listening to music on KSL Radio, hoping to hear my favorite song, "Blue Moon."

Nanie didn't live long after moving into Aunt Mary's home. I have only one memory of her death. Having not seen Nanie for several months, I was taken into Aunt Mary's bedroom where Nanie was lying on the bed under the covers. She seemed to have grown smaller than me, and the skin on her face, taught across her square jaw, was the color of sulfur. I knew I couldn't kiss or hug her, but Aunt Mary said I could touch her hand. Truthfully, I wasn't sure whose hand I was touching. It in no way resembled the hand that I had so loved and trusted.

My Nanie died on March 28, 1954, one day after the thirteenth wedding anniversary of my parents. After her funeral, I put a story she had written for me into my Mutual Improvement Association (MIA) "Treasurers of Truth" scrapbook. I glued blue, black, and silver rickrack around the edge of the paper to give it a special look.

To me Nanie had been soft snuggles, warm sugar cookies, belly laughs over my jokes and mistakes, and endless handmade dresses. She spent her whole grandmotherhood sewing on an old Singer treadle sewing machine for her ten granddaughters. I never wanted for a new dress for my birthday, the first day of school, Christmas, or Easter. Her dresses were made with deep darts and six-inch hems so they could be let out again and again.

Later that year, having been introduced to the concept of lay-away at Wickel's Men's Store, I decided to start my Christmas shopping early so that I could buy something special for my parents. Their bed was covered by the glamorous ice blue satin comforter, which still had an oversized "Do Not Remove Under Penalty of Law" tag disconcertingly displayed at the top corner. However, the only source of lighting was a flat overhead ceiling fixture—I thought two bedside lamps would be perfect presents. (Looking back, I'm sure the lamps were an unconscious attempt to shed light on their sex life. Physical contact with my mother was rare so I imagined my father's love life, if he had one, was lived on a very narrow ledge where any missed step could plunge him into perpetual abstinence.)

"Christmas Lights" was the name of my secret plan. I found two lamps with cut glass bases and ivory shades in the Sears Roebuck Catalogue, making a down payment from my summer earnings. The balance was paid by Thanksgiving, but lacking a hiding place, I asked the clerk for permission to leave the lamps in lay-away.

The afternoon of Christmas Eve, I left my mother's Dream House on the pretense of taking my present to Sherry, but instead walked downtown to Sears Roebuck. The trek was so cold and icy, I decided I would catch a Cook Transportation bus home. At the lay-away counter I discovered the lamps were packed in two large cardboard boxes. I could hardly get my arms around the boxes, not because of the weight, but the bulk. I left the store awkwardly, trying to balance one box on each arm and made my way across the street to the bus stop.

As soon as I sat down, a passerby told me that Cook Transportation buses were not running because of the holiday. I would have to walk home after all. On the bus bench, I struggled to free the lamps from their boxes and tried to create three smaller packages—two for the lamp bases, the shades double-stacked into a third. Wrapping them with the stiff brown packing paper, I abandoned the empty boxes on the bench. Half-way home I lost my balance on a spot of black ice. Before attempting to stand up, I heard the glass shards rattling, but my gift-giving heart was quickly covered with a protective coating of denial to ease me into the magnitude of my loss.

As soon as I hid the broken lamps behind our Christmas tree, my mother came out of the kitchen asking where I had been for so long. Not waiting for an explanation, she told me I needed to stay with Pepa while she went out. Only then did I see that Pepa was sitting in his wheelchair in the dark room.

Pepa seemed to smile at me, so I went to his side and whispered in his ear, "It's a surprise for Mom and Daddy!"

As soon as my mother left, I unwrapped my secret packages on the living room carpet while Pepa watched. Seeing that the lamp bases were beyond repair, even for my father, who was well-known for performing miracles with glue, I dissolved into tears. The loss of this surprise was more than my giving heart could bear. I had worked so hard! I had been so clever and so careful. Now the chance to see my mother the way she had been the previous Christmas was lost!

Through my sobs, I told Pepa the whole long story of preparation and anticipation. He was clearly distraught, using his cane to try to motion how I might do something, although I could not imagine what. He began to cry in frustration and became extremely agitated. While trying to comfort Pepa, I heard my father's footsteps on the gravel drive and quickly stowed the broken lamps out of sight behind the tree. Both Pepa and I were without hankies and snuffling from our blubbering, making it one of those rare moments when my father noticed emotional distress in others.

"What's wrong, Carol?" Daddy asked while his eyes stayed glued to the face of his father-in-law.

"Oh, it's a secret, Daddy, a Christmas secret!" I said. "I didn't mean to upset Pepa. I'm going to get him a warm wash cloth!"(a remedy for tears introduced to me by Aunt Mary).

I was trying to calm Pepa with the warm cloth, when I suddenly remembered the Christmas Eve program at the Ward House. I was supposed to play "The Little Drummer Boy" on the piano while the Primary School kids marched into the chapel. I was forced to leave Daddy alone with a still-sobbing paralyzed Pepa.

Walking home after the Christmas Eve program, as I approached my mother's Dream House, our living room was dark, except for the lights on the Christmas tree, which, to my great surprise, were visible from the street. It was the first time I had ever seen the drapes on our front picture window open at night. As I opened the door, Bing Crosby was singing about a "White Christmas" on the Magnavox and my parents were side-by-side on the couch. They seemed to be holding hands, but I may have only been imagining their fingers were entwined. I sat down on the grey carpet by the tree until Bing Crosby finished his repertoire. Then the disappointment of the day caught up with me. Having missed dinner, I descended the stairs to my basement bed, too tired to feed my cramping stomach.

A few hours later, I awoke to the sound of a crash and my mother shouting, "Frank, call an ambulance! Quick! It's Dad!"

Soon a siren announced the approaching ambulance. With my knees pulled up to my chin on our amputated couch, I watched two men maneuver Pepa out the front door on a gurney. The ambulance remained in front of our house for a long time, its red lights flashing through the window as if scanning our Christmas tree and the pale grey walls for anything moving or broken.

After my parents left for the hospital, I went into their bedroom and tried to imagine how the lamps would have looked. Taking their ice blue satin comforter, I lay down in front of the Magnavox to write in my diary, while listening to the Mormon Tabernacle Choir sing "Joy to the World" and "Hark! The Herald Angels Sing" on KSL Radio.

When Daddy returned home alone at dawn, I was still telling my diary about my day. As soon as Daddy said Pepa was in a coma and not expected to recover, big tears starting falling. Between sobbing breaths, I was able to tell him how I had upset Pepa with the broken lamps so it was my fault that he was going to die.

Daddy was still attempting to assuage my guilty feelings, when Mom called— she wanted to come home to shower and telephone her siblings with the sad news. Daddy persuaded me to go back to bed before he left to pick her up.

When I woke the angle of the sunlight shining through the basement window told me it was late afternoon. It certainly didn't feel like Christmas Day. At first I thought I had wet the bed; then I feared it was me, instead of Pepa, who was dying. My sheets were wet with blood, not urine. The house was very quiet. I climbed the stairs slowly uncertain of whom or what I would find. I was very relieved to see Daddy napping on the amputated couch.

I blurted, "Something is wrong with me! I'm bleeding!"

Daddy told me what was happening to my body. I clearly remember him explaining, "You will be okay, Carol. It is perfectly natural, but I think it might be a good idea for you to take a shower. I will get your bedding into the washer. Then you had better go next door to ask Donna if you can borrow a few menstrual pads until tomorrow."

Borrowing the necessary pads from Donna was humiliating. Sitting at the kitchen table, while Daddy made me toast and cocoa, I wondered why my mother had no supply of pads. My first and last glimpse of one had been a visual aid in one of Elaine's life science lessons.

Daddy looked at me for several minutes before speaking. "I think for now that it might be best not to mention to your mother that you have started to menstruate. She has a lot on her plate right now, okay?"

"Yes, Daddy, I promise. I won't."

Pepa died four days later in Logan LDS Hospital. His death meant that Christmas was always an unusually stressful time for my mother and me. I still feel responsible for Pepa's death.

-17-

Bras and Cars

The star of sixth grade at Adams School was my teacher, Mr. Lewis Downs. Like most girls in my class, I had a dual focus all year—whatever Mr. Downs taught and whichever way his body moved from blackboard to desk to window. But then the day came when not only was I unable to hear what Mr. Downs was saying, I couldn't imagine ever being in his presence again.

Until that day, school had been one of my favorite safe havens. Our whole class played kick ball at recess with Mr. Downs as referee. I loved kickball, relay races, and jumping over hurdles. But safety ended suddenly on a Monday afternoon when I arrived home from school and heard my mother announce Mr. Downs had telephoned.

"Why? Did I do something wrong?" my voice squeaked.

"Well, Carol, he suggested it is past time for you to start wearing a bra," she reported. "Some of the boys are snickering and he thinks it would be best for everyone if you dressed more modestly!"

The blood ran upward from my groin and collected in my cheeks. I would never be able to face Mr. Downs or anyone in my class again! How could he have said such a thing? My Nanie was gone; the flow of her dresses had stopped; this was more than a dart that needed to be let out.

Thirty minutes later my mother drove me to Sears Roebuck, where I was fitted for my first bra. She bought several panty girdles for herself since we were in that department.

"It probably won't be too long before you need one of these too," she said, as the clerk stared at my body before ringing up the sale, making Sears Roebuck one more off-limits store.

In the morning my mother made me change three times until she judged the bra barely noticeable beneath the bodice of my dress.

"You are going to need some larger clothes," she said. "I will talk to your father tonight."

Leaving my mother inside her Dream House, I walked toward the Adams School playground. But instead of going inside, I veered into Adams Park abutting the north side of the school grounds and kept going. I did not go to school that day or for the remainder of the week. Each morning, taking my lunch and school bag, I made my way across town, past Temple Boulevard and down to the willows of Canyon Park. I spent hours writing in my diary, reading, and skipping rocks in the Logan River. One day when it rained, I took refuge in the Logan Public Library. By the end of the week I had finished four library books.

My defection worked until the following Monday. When I wasn't in class the second week, Mr. Downs telephoned my mother, expressing his concern that perhaps he had "upset her." Upset my mother! What about me? I was escorted back to school by my mother the next morning, but it was already too late. I was now under the influence of unstable levels of estrogen and progesterone.

First, I refused to participate in any recess games or other activities. Then Jerry Fuhriman asked me to go to the movies. Even though I didn't like him, I was so mad at the world I agreed in a gesture of irrational spite. Jerry kissed my cheek in the darkened theater and told everybody! Scott Olsen wouldn't look at me, let alone write me a note; Sherry stopped talking to me.

Desperate to escape these social punishments, I couldn't wait for the school year to end. Instead of Nielson family babysitting, I signed up for a summer of bean and cherry picking. This was part of my new secret plan to work so hard and earn so much money as an adolescent agricultural laborer, that I would be able to buy a bus ticket to the other side of the world. At the end of August, I would run away from home.

The school year ended and every morning I dragged myself out of bed before dawn, walking down to the Mormon Tabernacle grounds on Main Street to board a bus with other Mormon kids to harvest beans in the fields of Cache County. Later in the summer, the same bus took us to the periphery of Brigham City, near the Bear River Migratory Bird Refuge, where we climbed cherry trees to harvest the sweet fruit, fighting off gulls. For both crops, we were paid by the pound. At the end of the first week of bean picking, my mother said my earnings were less than the cost of my sack lunches. This was an effective challenge. After her economic criticism, most days I was the high earner.

In the middle of summer, I decided that instead of buying a bus ticket, I needed to save enough money to buy a car because a vehicle would not only take me a desirable distance from home, it would also provide a much needed place to sleep and eat.

Simultaneously, a yet-to-be-identified hormone launched me into a conflicted mechanical puberty. Overnight, I became car crazy. The 1939 Chrysler Imperial was an unending embarrassment. Incongruent with my planned departure, I began lobbying my father to buy a new car, developing an encyclopedic expertise in brands and models.

My father's oldest brother, my Uncle Bob, arrived from Chicago for a visit in his new 1954 two-tone pink and black Ford Fairlane. Instantly it became my favorite. I begged Uncle Bob to drive me past Scott Olsen's house in Hillcrest, hoping that Scott would see me in my cool dream car. Scott never did.

A third split persisted between my secret plan and day-to-day reality. Before his death, Pepa had presented me with his old clarinet. While riding on the harvest crew bus I taught myself how to play, which gave me another goal—trying out for the Logan Junior High Band. I imaged myself marching down Main Street, heavily disguised as a car-crazy male in a heavy green and white wool band uniform—my breasts smushed so flat beneath braid and buttons nobody would ever know I needed a bra.

The author in her Logan Junior High Band Uniform, September 1954

Gender bias meant I was one of only three girls chosen to be in the Logan Junior High Band. At the end of the first week of school, those chosen were instructed to come back after supper to be fitted for band uniforms. Sherry and I agreed to go together since, as one of the majorettes, she needed to be fitted for a twirly skirt. Although we never spoke about my prior infractions, apparently I had served my Sherry-imposed social sentence by our summer-long separation.

Over dinner, I asked Daddy a seemingly innocuous question about how the gearshift of a car worked. A few years later, he told me he thought my interest in cars was an attempt on my part to avoid admitting to myself I was becoming a woman. Maybe so. None of the other girls I knew were interested in cars. Whatever hormone was moving me in this direction, by the time we finished eating, Daddy had drawn a sketch of the Chrysler's gear shift on the blackboard in our kitchen. At that virgin-mechanical moment, I actually believed I had an adequate understanding of the function of the clutch, the brake, and the gas pedal, as well as the relationship of those three pedals to the five gears.

My parent's bridge club was gathering at the home of Kenneth and Fae Gottfredson that night. Daddy told me to call him after I returned home from the fitting. Also, he said Sherry could spend the night, a permission that was rarely granted. Then instead of walking the single block to the Gottfredson's house, my father and mother unexpectedly drove away in the Chrysler because my mother was wearing high heels and didn't want to walk even a block.

My mother's footwear put the first crimp in the dangerous, hormonally-driven plan I had conjured up to elevate Sherry's and my own seventh grade newbie status by borrowing the Chrysler to drive to Logan Junior High for the fittings. I was twelve years old and had never driven anything more complicated than my two-wheel bike. Had I been in my right mind, I would have forsaken the plan as soon as my parents unexpectedly took the Chrysler, but, no, I was cruising for a collision.

Taking my mother's car keys from the hook by the back door, Sherry and I made our way to our chariot—each struggling with a stack of *Life Magazine* to use as improvised booster seats. Before leaving the house we made a division of responsibility for the operation of the vehicle we were about to "borrow," having convinced ourselves we were not engaged in stealing because, after all, the car belonged to my parents. My role was to steer, while working the gas, brake, and clutch pedals. Sherry would do the shifting. My father's blackboard schemata had been carefully copied onto a small piece of paper to ensure that Sherry had the necessary guidance for her role.

Positioning ourselves in the Chrysler's front seat on top of unstable magazine piles, we could see the bridge club couples through the Gottfredson's front picture window. I don't recall worrying if they would look out the window and see us—another clue to the nature of the adolescent criminal mind. One measure of the degree of concentration required by bridge, is that, of the dozen adults sitting around three card tables, not one of them noticed Sherry and me climbing into the Chrysler. Blinded by bridge, the club members also failed to look up and notice our struggle to start the motor multiple times after it stalled in response to our virginal clutch coordination. Nor were they distracted by our repetitive efforts to maneuver the Chrysler out from between the cars of two other club couples.

Sherry and I finally lurched away from the curb in search of status. We continued to grind gears all the way to Logan Junior High School. The motor died as we were attempting to brake and park a safe distance away from the other three cars in the school parking lot.

Abandoning our chariot and brushing off our clumsiness like breadcrumbs, we coolly strode into the Logan Junior High gymnasium to join the other seventh, eighth, and ninth grade band members. I dangled my mother's car keys to symbolize our special status-linked power, while making serial offers to give anyone in need a ride home.

Their Heavenly Father must have been watching over them because absolutely everyone declined. Disappointed that we had no takers, Sherry and I sat on the front steps until the sun went down. We judged it best to make our way back to the Gottfredson's under the full cover of darkness. When the janitor's car finally pulled out of the lot, we got back to our uncoordinated double-driving, grinding gears all the way up Temple Boulevard, a street chosen because it had only one traffic light. As soon as we passed through the light without stopping, which Sherry thought was red, but which I argued was yellow, Sherry turned around to see the flashing lights of a Logan City Police car directly behind the Chrysler.

"Carol, you've got to stop!" Sherry screamed. "It's a policeman!"

For the first time, I adjusted the rear view mirror in order to see for myself, while attempting to mimic the cool equanimity of Ben Evans.

I told Sherry, "I don't think you are supposed to stop unless you hear a siren. They flash those kind of lights all the time."

We continued moving forward for a few more blocks with the flashing lights illuminating the inside of the Chrysler. Then the police car pulled alongside and the officer motioned for us to pull over. Even I saw that. I steered the Chrysler to the side of the street, while the police car pulled in front of us and stopped. My braking role not yet perfected, the heavy Chrysler rolled on, stopping only after it hit hard against the rear bumper of the police car. The sensation and sound of the two cars making contact caused me to wet myself.

The uniformed policeman came to my window, shining his flashlight into my eyes. "Young lady, have you been drinking alcohol by any chance?" he probed.

"Oh, no, sir!" I said. "Smell my breath!" as I leaned out the window and aimed my breath towards his face—greatly relieved that it was a day when I was not emitting icky burps.

"Well, I have never seen worse driving in my life! Let me see your license!" his voice was stern.

I fumbled around in the front seat, trying to spread my skirt so that my now damp stack of *Life Magazines* remained hidden.

"Oh, Officer, I'm so sorry!" I said. "I must have left my purse at home. I don't seem to have my license with me."

The policeman gave me a ticket and explained that I would have to appear in court in two weeks or I could clear the ticket by taking my license to the police station—adding the caveat that I would still have to pay a fine for driving without having my license with me.

"Please try to drive a little more carefully, Miss!" he admonished, before climbing back into his police car and driving away.

Sherry and I were too frightened to feel any sense of relief that we had not been taken directly to the Logan City Jail.

We restarted the motor on our third try. We now faced the greatest driving challenge of our short motoring career—parallel parking. We circled the bridge club block and positioned the Chrysler on the right side of the street, but parallel parking was beyond our current level of coordination.

I could see my parents through the front picture window, laughing and talking as if they hadn't a care in the world—a confusing sight that filled me with adolescent resentment.

In the end, I had to do something even less pleasant than breathing into the face of the policeman; I had to ask Jerry Fuhriman for help. His family lived around the corner from the Gottfredson's. In my damp skirt with small pieces of *Life Magazine* sticking to the back, I knocked on Jerry's door and made my humiliating request. Ronald Erickson was there on a sleep-over, so he came along too. After several attempts, I got the car in neutral. Then, with Sherry, Jerry and Ronald pushing, we managed to get the Chrysler back into the narrow space between the two others, although it remained a noticeable distance from the curb.

Sherry and I hardly spoke as we lugged our paper booster seats and what remained of our lives back to my mother's Dream House. I wanted to implement my original Greyhound bus plan, but all my money was in the bank and the bank wouldn't open until Monday morning. Sherry, demonstrating the true depth of her friendship, quickly decided that she couldn't spend the night after all, packed up her things, and went home.

My hands shaking, I finally forced myself to dial the Gottfredson's number. As previously instructed, I asked for Daddy. Unfortunately, Mom took my call. I wanted to hang up, but I had already identified myself to Mrs. Gottfredson.

My breath almost nonexistent, I whispered: "Mom, I've done something terrible! I think you better come home."

"Did you break a wedding goblet or a piece of my china?" she probed.

"No, nothing like that!" thinking this rare display of naiveté on her part was astounding.

"Alright, I'll come down when this hand ends!" she said and hung up without saying goodbye.

An eon or two later, I heard the Chrysler on the gravel. I couldn't imagine telling her, but I did.

"I'm not surprised," she said. "You haven't been yourself for months. I'm going back to the Gottfredson's. I'm not going to break your father's heart until we are on our way home."

Slamming the front door, she gunned the Chrysler hard, flinging gravel onto the lawn as she backed out.

I was wide awake when my parents returned. Nine months had passed since Pepa's death and I was once again sleeping in my old bedroom. My feelings of preoccupying guilt for having caused his death suddenly displaced by fear. I lay on my bed, breathless, waiting for my punishment.

Daddy came down the hall, but instead of turning into my room, he went into the bathroom. Exiting the bathroom, he didn't even glance in the direction of my open door or indicate the slightest interest in my condition. Then Mom passed by, also without showing any interest. Had I already managed to execute my run-away plan, my parents wouldn't have even noticed! Instead, they seemed to be invoking the silent shunning of my peers.

The hardest thing for Daddy to accept was the premeditated nature of my crime. Sitting across from him at the kitchen table the next morning as he examined my ticket, I had to face, not only him, but also his careful drawing of the gearshift on the blackboard.

Daddy said very little, but the blackboard diagram of the gearshift glared back at me, shouting: "Guilty! Guilty as charged! And with premeditation!"

My age and my crime eventually required that I go to Cache County Juvenile Court, an institution previously unknown to me, located in the basement of the Courthouse on Main Street. My father accompanied me and watched the judge put me on probation for a year and levied what was then a very hefty fifty dollar fine. I paid the penalty from my run-away savings.

My father never forgave the Logan Police Department. Years later he was still furious at the policeman for his stupidity—for believing me, for having let me continue to drive after what he had observed.

"She was lucky she wasn't killed or didn't kill somebody else!" he would say.

My mother always defended the policeman, "Carol always looked more mature than she really was, even when she was a little girl!"

The great car caper was the cherry on top of my demolished reputation. I wasn't surprised when Sherry didn't keep her mouth shut. Jerry Fuhriman and Ronald Erickson acted as collaborating witnesses to Sherry's barely believable incident report. Just like Vicky at the crayon box, Sherry escaped any penalty for her illegal shifting.

My social survival required a high level of pretense, so being alone was more relaxing than being with others. Plus, I had my diary and time to read more books than anyone else. Marlene Evans, who was in ninth grade, was in my homemaking class. For most of that year, she was my sole source of a friendly greeting.

Our homemaking teacher, Miss Helene June Funk, was a stickler for perfection. After I had sewn the placket in my three-yard cotton gathered skirt for the fourth time, Miss Funk made a careful examination of my work.

"The placket is perfect, Carol, but now there are too many stitch marks," Miss Funk said without smiling. Then while ripping the placket out in one quick lethal strike, she added, "You'll have to start over."

Marlene lent me her hanky and put her hand on my back while I tried to stop sobbing. Eventually I felt grateful for Miss Funk's strict teaching methods because she taught me to be a good seamstress. Having inherited Nanie's sacred Singer treadle sewing machine, I began making my own clothes.

As my feet worked the treadle in my old basement bedroom, I kept hearing my mother's warning on the day of my juvenile court hearing.

"Young lady," she said, "Unless you straighten up right now, you are headed for big trouble! If you keep on with your ways you will end up in the Utah State Prison spending the rest of your life sewing prison garments and guard uniforms!"

-18-

Love and Betrayal

The first time I thought about adulterous betrayal was in 1958 when my taciturn father expressed angry outrage at Elizabeth Taylor because she had "stolen" Eddie Fisher from his wife, Debbie Reynolds. My father was not a fan of anything except the New York Yankees. Movie stars were not part of his world. Then ignorant of both his family history and basic human behavior, I had no way of knowing that my father's uncharacteristic display of emotion was about the adulterous betrayal of his own father. Instead, I assumed he thought Debbie Reynolds was the better woman. Elizabeth was bad; Eddie Fisher was just a helpless pawn—yet one more version of the archetypal story of Adam and Eve.

Two years before Elizabeth Taylor stole Eddie, my father was offered a one-year faculty appointment at the American University in Beirut, Lebanon. Not only would his teaching stipend be an increase over his government salary, but American University would also pay the expense of moving us from Logan to Beirut. Knowing other USDA researchers had taken sabbaticals for similar purposes, my father submitted his request to the Bureau of Entomology in Washington for a year-long leave. It seemed like a very exciting prospect, but I was told not to speak about it to anyone until arrangements had been finalized.

Apparently considerable USDA agricultural talent was being drained away by universities and private companies. At least that was the stated reason for the denial of his request for leave to spend the year in Beirut. Not only was his request denied, my father was notified he was being transferred to a new research lab at Montana State University in Bozeman. There had been a bad outbreak of locusts on the eastern plains of Montana.

When my father showed my mother the USDA letter of denial and his transfer orders, she retreated to their bedroom, threw herself face-down on the ice blue satin comforter, and started wailing. She remained in retreat for three days, grief stricken at the impending loss of her Dream House. Seeing her sobbing into the satin, I remember thinking the reason ice blue was her favorite color was because she was always icy and perpetually blue.

My own feeling of loss was focused on Teddy Wilson, a boy I had met the previous winter at the ice skating rink. Every winter a low-lying corner of Logan's Central Park was manually flooded to produce a large outdoor rink. On Friday and Saturday evenings it was swarming with teenage skaters. I was in the warm-up shed when Teddy Wilson asked me to skate with him. Two years older, but not much taller than me, Teddy had dark curly hair and grease under his fingernails from working on his Model-A Ford. In retrospect, he bore a strong resemblance to Eddie Fisher. Saturated with adolescent hormones, I quickly accepted Teddy's invitation. After we had circled the rink a dozen times in silence, Teddy asked if I would like a ride home. Whenever I walked home after ice skating, my ankles ached so I gladly accepted.

Both skating and riding with Teddy were costly choices. Having broken the normative eighth-grade-girl-code by associating with an older, somewhat dangerous boy, the next day, as I noted in my diary, I was blistered by mean girl looks and blatantly passed notes. Teddy Wilson's father ran the State Loan Company at a time and in a place where borrowing money was close to criminal. Ignoring all social deterrents, I continued on my way. Teddy Wilson had already used his powerful pheromones to cast his spell over me.

For the remainder of the winter, Teddy and I went skating almost every Friday and Saturday night. When the ice melted, I began playing on my Ward softball team. Teddy was always there to give me an indirect ride home. I didn't keep Teddy a secret from my parents, but their concern was palpable every time I opened the front door, closely followed by the sound of Teddy's Model-A backfiring as it rumbled away. Understandably, the first thing my father required was that Teddy show him his driver's license.

After about twenty rides home, Teddy and I began occasionally going to the new drive-in movie theater when I was supposed to have been at a Mutual Improvement Association (MIA) meeting at the Ward House. The drive-in movie was where Teddy French-kissed me, six months after that first ride home. I am certain Teddy found the progression of our physical relationship glacial (as in before global warming), but I was trying hard to follow all of the Mormon admonitions about chastity. After that kiss, I was terrified his tongue had made me pregnant.

My father had instructed me to keep the American University offer a secret until his year-long leave was approved, so I hadn't told Teddy about the possibility I would be leaving Logan in the fall to live half way across the world in Beirut, Lebanon. When the destination changed to Bozeman, Montana, I waited to tell him until the last day of school, the same day the "For Sale" sign was tacked up on the front of my mother's Dream House. I sensed Teddy withdrawing even as I spoke.

Despite Elaine's early life-sciences instruction about how Blackie and Snowball became mothers, I was remarkably ignorant about my own body. I had begun to worry about being pregnant because for most of the summer I had a kind of dull ache in my belly. On a night in early July, I went to bed early with a temperature and abdominal pain. Soon I was on the bathroom floor violently vomiting. At the hospital, Dr. Omar Budge announced that my appendix had burst and he had to operate immediately.

In the recovery room, still impaired by anesthesia, I began begging my mother. "Mom, please don't make Teddy mad at me!"

This was precognition. Teddy came to see me at the hospital only once. It was awkward. I was not the same. I was missing parts and there was a big red scar the full length of my abdomen. Plus, I was moving away; I was definitely damaged goods. Teddy was already moving on.

When I recovered enough to travel and my mother's Dream House was under contract, we drove to Bozeman in search of a new place to live. There was nothing, not even an empty Quonset hut. Bozeman seemed as short of housing as Logan had been right after the war. We put our name on various waiting lists and placed an ad in the *Bozeman Daily Chronicle*. My mother dripped a lake full of salty tears all the way back to Utah.

We drove into Cache Valley early in the evening and, after carrying our suitcases inside, we split into two task groups. My father started watering the grand garden, soon to belong to someone else, while my mother got back in the car to drive to Steed's Dairy on Main Street for milk and bread. Surprising both of us, I jumped into the backseat to accompany her.

While my mother was inside Steed's, I kneeled on the back seat to watch the cars cruising up and down Logan's Main Street through the high back window. Within minutes Teddy's Model-A passed and I thought I saw a girl's head on his shoulder. I begged my mother to follow Teddy, even though I was already slipping into denial, deciding I must have imagined the female head. Surely, Teddy couldn't have already betrayed me! I had only been gone five days!

Maybe my mother agreed to follow Teddy because she frequently felt followed herself. Whatever the motivation, this tailing of Teddy is the only time I recall my mother and I cooperating as co-conspirators. My heart pounding, we caught up with the Model-A on Main Street just before Teddy turned onto Temple Boulevard. My mother confirmed I wasn't hallucinating—someone was sitting in my favorite place! The Model A continued past the University campus before turning into Hillcrest, the new housing development at the mouth of Logan Canyon where Scott Olsen lived. When Teddy slowed, then stopped, I pressured my mother to circle the block.

The second time we passed his Model-A, Teddy was kissing the girl. I looked directly into her eyes as she looked past Teddy at me. It was Elizabeth Budge, a dark-haired, full-breasted temptress. Her father was the first cousin of the surgeon who had just deflowered my belly. Three years older and far more sophisticated than me, she had stolen Teddy! When I caught my breath, I told my mother to drive home as tears poured down my cheeks. I felt helpless, powerless, shamed, and enraged in the face of the evil Elizabeth.

Teddy's car was already parked in front of our house when my mother turned into our driveway. Seeing his car, my hopes rose. I thought that Teddy had come to tell me Elizabeth had forced him to give her a ride and he was still mine! My mother drove the Chrysler into our garage and disappeared through the side door. I was left to face Teddy alone.

Teddy got out of his car and stood on the sidewalk, wearing a pink and white striped shirt, the short sleeves rolled up to highlight his tanned biceps. My heart was pounding; my throat closing.

"What in the hell do you think you are doing?!" he shouted.

"I wanted to see who was in the car with you!" I sobbed.

"Well, don't ever pull a stunt like that again!" Teddy said, grabbing my upper arm. "You hear me?"

Pulling away, I sat down on our front lawn, putting my head between my legs in an effort to wipe the mucous from my nose on my jeans since I didn't have one of Aunt Mary's hankies in my pocket.

"I'm sorry, Teddy. I didn't mean to upset you. Just tell me who she is, please, Teddy," I begged.

Even though I knew exactly who Elizabeth Budge was, I didn't want Teddy to know I knew. What I really wanted him to say was that it was just a mistake. He didn't like Elizabeth Budge at all. She had kissed him. He liked only me. Wouldn't I please forgive him?

What he said instead was, "It isn't any of your business! You are leaving, so who I see is up to me! It's a secret! Nobody is supposed to know!" He spat on the sidewalk and turned to get back into his car.

I reached for his hand. "Please, Teddy, don't do this! Don't leave feeling this way! Please!"

Stiffly pulling his hand out of mine, Teddy climbed back into his car before I could think of another desperate plea in the night court of love. Revving up the motor, he sped away, back firing all the way past the Adams School.

Sick from betrayal and loneliness my last weeks in Logan, I rode my bike down to the drive-in movie on Friday and Saturday nights searching for Teddy's Model-A and watched even though I had already knew the plot.

This was my first encounter with the Green-Eyed Monster and it had me completely in its grip. Mumbling to myself, I verbally sliced Elizabeth into little pieces with the kind of surgical precision that would have been the envy of her uncle. All my rage was directed at her, the other woman, not at Teddy. Now I knew other females could enter my territory in an attempt to take my mate. No matter how pure my devotion or how much I gave of myself, any male could be lured away by a more powerful female.

I thought of Elaine, but could find no remnant memory of her having ever told me about this kind of competition—only that male cats often killed their own sons to protect their territory.

No matter, whatever Teddy and I had was now contaminated—the simple sweetness lost forever. We had been so close. Now the distance between us could not be bridged. Blinded by betrayal, I couldn't see Teddy's low loyalty and sudden lack of tenderness for what it was. Instead, I blamed only Elizabeth, whose seductive power had soiled the sacred space between Teddy and me.

On Labor Day, I rode out of Logan to start life over in Bozeman. Just past Henry's Lake at the top of Targhee Pass, where a sign marked Idaho's border with Montana, my father expertly pulled the Chrysler onto the shoulder of the two-lane highway and stopped. In a gracious gesture of parental forgiveness, he instructed me to take his place behind the wheel. A month later, I was living comfortably with my car hormones and a legal license to drive as a fourteen-year-old resident of the State of Montana.[13]

My father's full parental pardon imbued me with critical courage as I moved into a new and strangely secular culture—leaving Teddy Wilson and Mormonism behind.

-19-

Dreaming of Being Mrs. J.C. Penny

In Bozeman, I was determined to live on the right side of the tracks. I didn't want to be judged a "bad girl" ever again. Fortunately, being seen as a "good girl" was easy because Gallatin County High School was still part of the old Wild West. Students smoked on the school grounds and openly drank beer at school dances.

The first mark on my clean slate was an unplanned name change. When roll was called in my home room, the teacher used my full given name, "Joan Carol." My new peers quickly shortened this to "J.C." while teasing me about my "strange accent." I did speak with a different vernacular. In Bozeman, I was still an "O" in a group of "X's"[14]—but I felt more like a cattle dog living in the midst of a large herd of free-roaming buffalo, compared to Logan where I had felt like a wild animal trying to hide among corralled dairy cows.

There was another freshman newcomer to Gallatin County High School that year. His name was Barton Penny and he was the first boy I ever dreamed of marrying. Because of my new nickname, I sometimes imagined myself carrying nomenclature that closely matched that of the famous retailer, "Mrs. J.C. Penny."

Having entered Gallatin County High School with blank social slates, both Bart and I benefited from the kind of temporary popularity that comes with having yet to earn enmity or envy. Bart was elected class president; I naively agreed to become a cheer leader.

Bart had moved from Butte to Bozeman following his father's remarriage. His biological mother was mysteriously missing from his life and he was living with his older brother and his sister-in-law, both students at Montana State University.

My mother's response to Bart from their first encounter was extraordinarily positive. At their first meeting, he began addressing her as "Beautiful." To my great surprise, she was far from offended. Now, when I look back on their relationship, I realize Bart was the perfect substitute for the son my mother had so wanted, but lost.

In many ways, the four of us were all migrants living in temporary camps. We had been forced to revert to apartment living; Bart was staying in a narrow room in the apartment of his brother and sister-in-law. The only shelter we had been able to find was a small two-bedroom apartment on the second floor of a tenement-like complex called the Townhouse Apartments. With its substandard construction, paper-thin walls, and lack of charm, it was the perfect place for my mother to mourn the loss of her Dream House.

Bart Penny, Bozeman, Montana, 1956

The 1939 Chrysler Imperial reached the end of its long life, forcing my father to purchase a new car with cash and caution. He chose a 1957 cream-colored four-door Plymouth Belvedere sedan with a long coral strip along each side ending in a giant fin. It was part fish, part car! The seats were covered with cream-colored Naugahyde and the roof was lined in black fabric imprinted with little gold stars.

The Fish-Car—1957 Plymouth Belvedere

Because my mother was extremely restless and unhappy and the Fish-Car was a much more comfortable place to spend time than the Townhouse Apartments, we began treating her Bozeman blues with weekend car trips. Beginning with the first finned excursion, Bart was invited to ride along.

We drove south through the Gallatin Canyon toward West Yellowstone, sometimes going into the park, although my mother always resisted. Photographs from that year show her standing in snow in West Yellowstone wearing her high heels, a form of fashion insurance protecting her from engaging in any activities that might possibly be bear-related. Bart, my father, and I fished the Gallatin, the Firehole, and the Madison Rivers, laughing as we stumbled into sinkholes along the banks, while my mother sat in the Fish-Car reading books with covers made of brown butcher paper to prevent the FBI from using long-distance lenses to photograph the titles.

The four of us explored the major towns of Montana and many minor ones as well. We traveled to Butte, a great hole in the earth, to see where Bart had grown up. We drove to Billings for a rodeo and fair, to Helena for Christmas shopping, to Missoula to tour the University of Montana and to Glacier National Park for the pristine lakes, and high mountain vistas. When the weather was terrible, we drove thirty miles east to Livingston, shared a restaurant meal, turned around, and drove back to Bozeman.

In recovery from female betrayal, I kept my distance from girls in Bozeman. Instead, Bart became my boyfriend, my best friend, and my trusted confidant— like a twin brother. I used to play our piano, while Bart sang "Day-O, The Banana Boat Song," As he sang, standing behind me with his hands on my shoulders I was surprised to find myself wishing he would drop his hands from my shoulders to my breasts.

Margaret's high heels as bear protection, West Yellowstone, Montana, February 1957

September 1956 to August 1957 was one of the sweetest years of my life. It was also the beginning of my rock and roll experience. Music took control of my mind and body. Fats Domino sang "On Blueberry Hill" while Bart and I danced our hearts out in the Elks Club above the Fire Station on Main Street. We also loved Jim Lowe's "What's Behind the Green Door" and the music of Elvis Presley, James Brown, and the Platters. It was a wonderful year of dancing and gentle touching.

Ten months after our arrival in Bozeman, my father was notified that he was being transferred again. This time the destination was Bakersfield, California, at the southern end of the San Joaquin Valley in Kern County. His new assignment was to lead a "war on the spotted alfalfa aphid" (*Therioaphis maculate*) then decimating alfalfa fields across the West.

When the Mayflower Moving man parked his giant semi-truck outside the Townhouse Apartments at the beginning of August, he hired Bart to help him load. Bart was delighted to be hired for the then-unheard-of wage of three dollars an hour.

The previous night, Bart and I had lain together on the bed of his brother and sister-in-law. We were alone in their apartment, except for Bart's pet barn owl, Olly. Having rescued Olly from near death, the way Bart cared for that owl was a big part of why I fell in love with him. I would have made love to Bart that last night in Bozeman if I had more bodily courage. But I was menstruating—both ashamed and afraid.

Maybe those moments felt so magical because that night was the first time I had ever laid on a bed with anyone except my cousin Julie Ann. It seemed the radio played only meaningful songs. Those hours had an unbelievable softness—the cotton quilt, the nightlight in the bathroom, the music, and the yellow eyes of Olly watching us from his perch. We shared a beer from his brother's refrigerator, my first.

Bart Penny with Olly, his rescued barn owl, Bozeman, 1957.

I knew the wonderful things I had done with Bart would never happen again. Already grief stricken, I was anticipating the loss of the instant safety I felt the first time Bart held my hand, making me feel I really belonged to someone. He had my full trust as we shared secret confessions on cold nights walking home. I loved our joyful dancing until we were fully satiated. Bart had bought me my first Coca-Cola, a drink prohibited by the Mormon Word of Wisdom, instructing the young man behind the counter of the soda fountain in the drugstore on Bozeman's Main Street to garnish the Coca-Cola with a slice of lemon. Nothing has tasted as good since.

We had spent hours sledding on a slope we called "Blueberry Hill." There was the back seat snuggling in the Fish-Car, interrupted by bouts of wild playfulness in the great outdoors of Montana. And then finally there was our last warm summer night together when I experienced the sensation of Bart's hands on my breasts.

Even my mother gave Bart a hug as we bid him goodbye outside the Townhouse Apartments. He looked utterly forlorn, rubbing his hands over a twisted knobby branch from a contorted lodge pole pine (*Pinus contorta*) my mother had given him while packing. She had picked the branch up on the north side of Yellowstone Lake during one of our expeditions into the Park. The branch had been bleached almost white by water and sun. As we drove out of the park that day, she kept repeating that the branch reminded her of an angel's wing. Only Bart had agreed with her.

As the Fish-Car pulled away, Bart raised the wing-shaped branch over the top of his head and began running after us, shouting, "I'll make a lamp out of it, Beautiful, and send it to you!"

-20-

A Female Twin

It was one hundred ten degrees when we drove into Bakersfield on Interstate 5. Unfortunately, our new Fish-Car was not equipped with air conditioning and our collective coping skills were low. I was lovesick for Bart; my mother was paranoid and prostrate from the heat; my father was preoccupied with the challenges of his new research assignment. Our common complaint was that we had no place to live.

Over the next two weeks we spent a great deal of time in a Carnation Ice Cream Shoppe trying to recover from frequent forays into the steaming Bakersfield real estate market. I was shocked that Bakersfield had not one, but five high schools. The place resembled a sprawling weed desperately putting its tentacles into the soil of the surrounding oilfields. For the first time in my life, I was going to have to learn how to live in a city, instead of a small town.

After a frustrating search, my parents compromised and purchased a new tract house in East Bakersfield on Hollins Street near the bluffs and Bakersfield Junior College. Quality-wise, this poorly constructed cookie-cutter structure was about half way between the cardboard tenement in Bozeman and my mother's Dream House in Logan.

The Hollins Street location meant that I would attend East Bakersfield High School (EBHS), joining a student body of twenty-five hundred students. The size made me nervous, but I also sensed that EBHS might provide me with total anonymity, a virgin-like experience.

Mourning Bart's comforting companionship, I stepped into the halls of EBHS not knowing a single soul. The sophomore class of about five hundred students was far more diverse than either Logan or Bozeman—about twenty percent were Hispanic and Black, two dozen were Mormons, and another ten percent were Jewish.

After learning my surname, a Jewish student asked me, "Who did your nose job?"

For our first California Christmas, Bart traveled from Bozeman to Bakersfield on the Greyhound Bus to spend the holidays with us. He arrived in time to escort me to the EBHS Christmas Ball; by the time he boarded the bus back to Bozeman on New Year's Day, I had regressed into an acute state of longing for his presence.

On the second day of the spring semester, I met Nonny Thomas.* She was struggling to get the neck of a long-sleeved black leotard past her Rubenesque buttocks in the girls' locker room. My gym locker was next to Nonny's, so I couldn't open mine until she closed hers. I followed Nonny into the dance room where she watched herself in the mirror, as if hypnotized, lumbering through a self-choreographed solo modern dance routine to Eddie Fisher singing "Oh, My Papa." After class, Nonny invited me to go with her to the modern dance concert at Bakersfield High School, a rival in all competitions. Their troupe of dancers was infinitely better than ours; that shared critical judgment was the beginning of our friendship.

Nonny Thomas, East Bakersfield High School, 1958

---

* I only learned that my father's full given name was Francis Valentine after I was elected "Queen of Hearts" at the "Valentine Valtz" in February. That event, led to his confession. When I told Nonny about my father's real given name, she said that her full given name was "Naneen Francis." Coincidentally, the first time I heard her name during gym roll call, I thought the teacher said "Nanie Thomas" which I took as a good omen.

Nonny's mother was a divorced second grade school teacher. Her father had disappeared from their family life when Nonny was eight. Their household was casual and female-oriented, except for the irregular appearances of Nonny's brother, who had already graduated from EBHS and spent most of his time with friends in sports bars.

Nonny struggled with her studies, not because she wasn't smart, but because she lacked interest in the curriculum. She once resorted to cheating in order to avoid being disqualified from a cheerleading contest because of grades. Afterwards she wrote me a long confessional note. She recalled my scolding her, but I only remember feeling that Nonny wasn't meant to have to suffer through math and biology. Already an accomplished artist and dancer, I thought Nonny should have only been required to draw and dance.

Nonny's most important goal was to be elected varsity cheerleader. She had already been elected cheer leader for the sophomore class. But since teams rather than individuals campaigned for the varsity-level laurel, Nonny coerced me and a shy blond named Dona Herr to join her. I had been completely cured of cheerleading in Bozeman. At Gallatin County High School I had learned how uncomfortable I felt in short clothing trying to move in sync with others, while simultaneously exhorting an audience. Nonetheless, in my over-responsiveness, to any of Nonny's needs or wants, I reluctantly agreed.

Nonny designed our outfits with feminine delicacy in mind, collaborating closely with her mother's dressmaker to bring puffed sleeved blouses with Peter Pan collars and pleated skirts to life, while hounding Dona and me through endless hours of practice on a body torturing routine.

Our campaign speeches had to be delivered on stage in front of the full student body. I insisted that we speak from memory rather than holding note cards, which I judged too amateurish. Nonny hesitantly agreed, fearful she would forget her lines.

On the day of the election, our opposing team arrived at school wearing sporty striped sweaters with an outsized "E" sewn on the front, each carrying a giant rag doll made in their image, which they used to make their routine charming and entertaining. The instant I caught sight of them, I knew we had lost. Worse yet, I was the one who fell mute before the full student body.

After we lost by fifty votes, Nonny, suffering from the heartbreaking defeat, stayed in bed for a week. In her effort to avoid conflict, she said the loss wasn't my fault, but I knew otherwise. Nonny had to settle for being a senior class yell leader.

For me, being with Nonny was like a perpetual ticket to a good romantic comedy. Her total engrossment in whatever captured her attention was endearing. She was high quality human entertainment—loveable for her losses, as much as for her wins, for her ups, as well as her downs. For the remainder of my high school education, Naneen Francis Thomas took the place of my comforting and reliable diary. The only thing I kept secret from her was the condition of my mother's mind.

-21-

Speaking Truth to Power

The war on the spotted alfalfa aphid drew my father into a new area of entomology, one that mixed bugs with botany. Leaving DDT and Dieldrin[15] behind. Frank V. Lieberman was tasked with attempting to breed a resistant strain of alfalfa. He had experimental test fields as far away as the Lahontan Valley in Nevada. Even on those days when he wasn't traveling, my father left our Hollins Street house before dawn, returning after dark.

Physical relocation, even the painful transition to Bozeman, seemed to provide my mother with a temporary reprieve from her schizophrenic demons. This was true for her first year in Bakersfield as well. My mother's functionality and behavior was so improved she was able to obtain and hold a secretarial position with the California Youth Authority. As a member of a too-closely-supervised typing pool, she seemed less alone, more supported in her human interactions, although I never knew her to have a friend.

Between my studies, extracurricular activities, and a part-time sales job at Sears Roebuck in the Ladies' Foundation Department, (where I obtained my mother's favorite panty girdles on my employee discount), my family's first year in Bakersfield was a kind of working blitzkrieg.

On the day after my mother's birthday, she read an article in the *Los Angeles Times* about Rose Marie Reid, a Mormon and a successful California swimsuit maker. After Rose Marie's niece married a Jewish man, Rose Marie began advocating for the Mormon Church to increase its efforts to convert Jews. Apparently Rose Marie had said in print what God had been whispering in my mother's ears for years. Bingo! My mother quit her job and drove to Los Angeles the next morning after Rose Marie Reid innocently agreed to a meeting.

I was not surprised when, after their initial meeting, Rose Marie was always too busy to meet with my mother again. After several rebuffs, my mother decided to tell the Mormon bishop in Bakersfield what God wanted done.

The response of the bishop was, "God does not reveal his plans to women; he only does so to Mormon males, and only those worthy enough to become members of the priesthood."

Undeterred, my mother moved up the Mormon hierarchy. By the time she reached the Quorum of the Twelve Apostles[16] in Salt Lake City, her two brothers-in-law, both bishops of their respective Wards in Salt Lake City and Bountiful, had been alerted. Her sisters and brothers-in-law maintained that Margaret's missionary zeal was the result of having married a Gentile. First, her sisters were asked to dissuade her, then their priestly husbands, but all efforts failed. Margaret Audrey Beck Lieberman was adamant that God had told her this work was to be done and it was her deepest obligation to see that it happened.

The more my mother persisted, the more her Mormon family members and Church authorities increased their pressure. After a few months of back and forth, Church authorities threatened her with excommunication—a strange response to a thirty-nine year old woman who hadn't been inside a Mormon Church since she was fifteen.

The threats pushed my mother into a state of increasingly pronounced paranoia. She became convinced that the woman who lived across the street was an FBI agent assigned to keep her under surveillance and was watching our house through not-quite-closed Venetian blinds. Soon thereafter my mother stopped leaving our house, keeping both our blinds and drapes closed as a double protection against surveillance.

There was a large food mart where the bus I used stopped a block from our house, so I began bringing a few groceries home after school or work. In retrospect, these food forays were my first efforts to assume my mother's responsibilities whenever her paranoia was particularly paralyzing. My father seemed blind, as if his childhood had given him immunity to human distress. Instead, his antennae seemed to pick up non-human structural flaws; the slightest crack in cement, a tiny scratch in paint, a thirsty garden plant, or other sign of household decay caught his attention right away, as did any pothole or design flaw in public infrastructure.

My father's failure to notice his wife's obvious decline was also obscured by the fastidious attention she paid to her personal appearance. Unlike most suffering from serious mental illness, my mother washed and set her hair daily, pulled on a clean panty girdle before carefully choosing a dress for the day and applying Revlon Red lipstick. She was always perfectly presentable to any friend or foe who came to our door.

Before my mother became obsessed with Rose Marie Reid, I had been selected by Brock's Department Store to be EBHS's "Campus Deb" sales representative. At Brock's I was assigned to work in the Bridal Department under the supervision of Mrs. Mildred Verhaag. Then forty-five, Mildred was tall and slender, with an exquisitely chiseled face; her thick grey hair was cut in a modern, perm-free bob. Mildred believed girdles were the invention of a sadist and refused to wear one. After my first week of work at Brock's, she took me into the bridal dressing room to show me her French stockings and garter belt—items she swore were more comfortable and attractive than a girdle. This intimate transaction was triggered by my use of my Brock's employee discount for the purchase a flower-print latex panty girdle, not for my mother, but for myself. Mildred strongly suggested I return it immediately.

I loved working for Mildred Verhaag, helping brides into their dresses, while listening to her counsel them on make-up and wedding mores. Whenever we weren't busy, she freely coached me on my posture, makeup, hair, and wardrobe. Mildred Verhaag looked at me, saw I needed help, and generously gave me a taste of the kind of mothering I had never known.

My family role changed forever one Saturday early in October 1958 after a day filled with Mildred's careful coaching. As I stepped off the bus and began walking north on Hollins Street, I saw my mother in our front yard taping big pieces of cardboard over our front windows.

Reaching the curb in front of our house, I shouted, "Mom! What are you doing?"

"You know what I am doing and why!" she snapped back, her mouth tense with paranoid fear.

Shaking my head in adolescent disgust, I went inside. My father was asleep in his recliner in the family room, soothed by the sounds of late season baseball. Having just experienced some desperately needed mothering from Mildred, I suddenly felt unwilling to continue our family charade. In less than an hour, Dick Alton, an EBHS classmate, was due to arrive to pick me up for our first date. I felt angry. Why wasn't my father paying attention? Things were getting worse! Compared to how I felt in Mildred's presence, my situation at home was completely unbearable! Potential social shame stimulated my survival instincts. It was one thing not to get any meals, nurturing, or social coaching—it was quite another to suffer a humiliating display of our family secret in front of my peers.

I sat down on the footstool, calling my father out of his nap, "Daddy, Daddy, please wake up!"

My father opened his eyes and smiled at me, "What time is it? Are you already home from work?"

"Yes, Daddy. I need to talk to you! Do you know what Mom is doing out front? She's putting cardboard over our windows so that the woman across the street can't watch her. Daddy, Mom is really sick! She needs help and we have got to get it for her!"

I had crossed the parental boundary and stood hopefully just inside the border staring at the guard. I desperately needed him to help me find my way through what was extremely treacherous territory.

Slowly he rose from his recliner before turning away from me and moving toward the glass patio doors. He stood silently surveying the gardenias struggling to survive next to a cement slab covered by a sheet of rippled polycarbonate roofing. As was his habit, he tucked his thumbs between the waistband of his pants and his belt, from which a metal clip, thick with keys, always hung. After a few minutes of silent visual contemplation, my father turned, walked past me into the kitchen, where he put the tea kettle on the stove.

I followed behind, begging, "Daddy, I need you to say something to me! Please, I can't stand living like this anymore. Mom is really crazy! Don't you know what I am talking about?"

My father remained mute. He waited for his tea water to boil and, only after he had placed a Lipton bag in his cup, did he turn to face me. "No, I don't know what you are talking about and I wouldn't know what to do if I did!"

My face began burning—then tears started to fall in big drops down hot cheeks. I felt simultaneously enraged and humiliated. There was no helpful guard, no rescuer, no protection, not even hope! Feeling as if I had been punched in the stomach, I quickly went into my bedroom, shutting the door behind me.

It was many years before I came to understand and forgive my father for his failure to respond to my plea for help. In 1958 denial was his only defense. It was the middle of the long era when it was our common cultural practice to lock up the mentally ill and to keep cancer diagnoses secret, even from patients.

But perhaps more importantly, although I didn't know it that October evening in 1958, to overcome the sins of his father, my father had taken a private vow never to abandon his family. These two factors—one driven by the larger societal context, the other by painful parental imprinting—kept my father at a blind distance from the implications of his wife's behavior. I felt that he had handed the responsibility for our family secret to me. Not only did he not help me find my way, he never left the heavily defended territory of denial.

Sitting on my bed in the dark, I attempted to quell my tearful trembling by holding my pillow tight against my abdomen. At the sound of the doorbell, adrenaline flooded my system. I had completely forgotten about my date with Dick Alton. I listened as my father opened the door and invited Dick inside our darkened living room, followed by the sound of the switches on the twin table lamps guarding our amputated couch.

Knocking on my hollow-core bedroom door, my father said, "There is a young man here to see you, Carol."

"Okay," I replied, "I'll be out in a few minutes."

I was still in my Brock's Campus Deb outfit, a lavender sweater and pleated skirt—an outfit that Mildred had pointed out a few hours earlier was not the most flattering line or color for me. I had no time to change; worse I had no idea about my mother's current location. I made a quick dash into our single bathroom, washed my swollen face with cold water, and applied some of my mother's Revlon Red lipstick. My own lipstick was inside my purse in the family room. I went into the living room to greet Dick, praying that my mother had disappeared off the face of the earth. Relieved when she wasn't there, I went into the family room to retrieve my purse, and saw her sitting in the dark.

"Hi, Mom," I said, "I'm going out to a movie. I should be home by eleven." The transaction sounded deceptively normal.

Walking down our front steps, Dick queried, "Are you guys getting ready to paint your house? Isn't it brand new?"

"I'm not sure," I said. "It's something my mother is doing."

I doubt Dick even noticed my swollen face. His eyes were focused lower, on the bust line of my sweater. My eyes, feet, and head ached in concert; all I wanted to do was to cancel and crawl into my bed. At the movie theater, there was a long line. Just as we reached the ticket window, Dick discovered he had forgotten his wallet. I told him that there was no problem because I had plenty of money. This first financial transaction was a harbinger of our future financial relationship.

After the movie, I asked Dick to take me home.

"But don't you want to get something to eat?" Dick seemed slightly puzzled until he remembered his financial dilemma. "Oh, I understand," he said, "We can get something to eat next time."

"It's not about the money, Dick. I'm just really tired," I explained, honestly attempting to assure him.

That night my sleep was episodic, disturbed by bad dreams, but I knew that there was no turning back. I was no longer able or willing to pretend there was nothing wrong with my mother.

The next morning, before going into the kitchen, I took a long shower, dried my short hair, and dressed as if going to work for Mildred. I had made up my mind to speak the truth again, even though I fully expected my words would be ignored. I was filled with a strange combination of congruency and fearlessness, the feeling of relief that comes with telling the truth.

My father was working at the kitchen table with a petri dish and tweezers, wearing his magnifying headgear, carefully counting the contents of a container of aphids from a field sample before placing them in a petri dish filled with agar.

My mother was washing dishes, her eyes fixed on the garage wall of the house next to ours—fortunately it was a wall without windows. In the living room, where we were still protected from prying eyes, it was as dark as night. I told my parents I had something important to say.

"Mom, would you come sit at the table?" I patted the worn yellow Naugahyde on her usual seat and she slowly dried her hands before sitting down. "Mom and Daddy, I tried to talk to Daddy yesterday, but I don't think he fully understood me. Mom, I believe that you have a mental illness that is affecting the way you think about things. You seem to be feeling more fearful and anxious every day. I am very worried about you. Tomorrow I am going to try to find a doctor who can help you. If you refuse to see a doctor, then I am going to find one to tell me what to do because I can no longer make myself pretend that nothing is wrong."

My declaration drastically disturbed the distance between the three of us. There was total silence for several minutes until my father ripped off his headgear and started pounding his fist on the Formica table top, disrupting his tedious count.

"There is nothing that can be done and you know it! You are just going to stir up a hornet's nest! I've already told your mother the cardboard is coming down. Don't make things any harder than they already are!" his voice was unusually loud—surprising because he rarely showed his anger.

My father stood, walked to the glass patio doors, and took his usual stance with thumbs tucked behind his belt, his eyes focused on terrain less than ten feet north.

My mother was silent, her eyes focused downward, as if memorizing the table's familiar Formica pattern. Suddenly she stood, then left the kitchen, heading toward their bedroom. I heard the bedroom door close, followed by the sound of the second door in their bedroom, the one which opened into the garage. Then I watched my father watch his wife back the Fish-Car out of the garage and disappear down the alley.

Although my words had done nothing except increase the already intolerable level of familial tension, something deep within me was no longer willing to pretend, ignore, look away, or smooth over the increasingly dysfunctional condition of my mother's mind. Our pseudo-normalcy was now naked.

Neither my father nor I said anything to each other for the remainder of the day. My mother didn't return until after dark. Hours earlier, my father had removed her cardboard window shields. Several weeks after this episode, my father agreed to hang bamboo screens from the gutters over the front windows to sooth my mother's paranoid feelings.

I was doing my homework at the dining room table when my mother opened the back door carrying a bag of groceries. I felt a confusing mixture of rage and relief. Like all impulsive adolescent heroism, my own was extremely short-lived. It was almost impossible for me to hold onto the sacred space I had created for myself in the midst of denial and madness. My mother's shifting symptoms were her most powerful distancing device. Every day a war raged inside my brain between suspicion and unobtainable confirmation: Was my mother in control of her behavior or had something else taken control of her brain?

My mothering deficit was so massive that the sight of her momentarily functioning in something resembling the role was very tempting. But I knew almost anything could be inside that grocery bag and I should have no expectation of being fed.

-22-

The Over-Achievement Defense

After my amateur impulsive intervention failed, I didn't even try to find a doctor to help my mother. Instead, as if I still in Logan, I found myself dreaming of ways to leave home again. Magically just such an opportunity presented itself when Mrs. Sallalee Ryan, an EBHS business teacher, sat down across from me in the library.

"Joan Carol, I brought some materials for you to review because I am hoping you will consider submitting an application to become an American Field Services (AFS) exchange student," Mrs. Ryan said. "I think your maturity and interest in science make you well-suited to the program. It would mean you would spend next year abroad."[17]

I stayed up all night completing the application. With my newborn-boldly-truthful persona, my response to the question on the AFS application about my religious affiliation was reckless, anything but mature. Here is what I wrote:

> *"My father practices no religion and my mother has recently been threatened with excommunication from the Mormon Church. While, I am still on the membership roll of the Church of Jesus Christ of Latter Day Saints, I never believed in Mormon doctrines and stopped active participation in 1956."*

When my father got up at dawn, I handed him the AFS application and told him to sign it, saying we could talk about it later. He signed without reading it and handed it back to me without a word.

The possibility of a long distance escape increased my tolerance for my father's denial. But I also made a more deliberate effort to minimize interaction with my mother—avoiding direct eye-contact as I moved between my bedroom, our single shared family bathroom, the always aphid-stocked refrigerator, and my life in the outside world. It was easier to maintain a blind distance if one had other demanding responsibilities. Just as my father's professional responsibilities kept him at a deniable distance from my mother's mental illness, I began using my own obligations as protection.

I worked at Brock's after school Wednesdays, Thursdays, and all day Saturday, the employment providing me with enough money to prevent hunger from pulling me home. Nonny and I were now full-fledged members of a mixed-gender social gang. There were always impromptu meals at hamburger joints, study groups, club meetings, dances, swim parties, must-see movies, or sporting events. It was easy for me to build a fortress of outside obligations.

Two teachers helped in my over-achievement defense. Six foot tall, Margaret Lucius Sprague, my geometry teacher, was a giant of a woman in every way. Mrs. Sprague's teaching was so inspiring, I labored over my assignments until they were perfect. The approval she gave me flowed effortlessly into my otherwise-empty emotional pockets.

Walter Evert Shore, my biology teacher, encouraged me to enter my honors research project on agricultural economics in the Kern County Science Fair. I agreed only because working in his laboratory during spring vacation was better than being at home with my mother.

After I won the County Fair, Mr. Shore hauled my project up to Fresno on Sunday for the week-long California State Science Fair. Fully intending to go to the awards ceremony, I had arranged with Mildred to be absent from work. But walking home from school, I remembered that the essay for the Daughters of American Revolution (DAR) essay was due the next day. I hadn't even started on a draft, plus I had a lot of other homework.

When Mr. Shore came to our front door to drive me to Fresno, I begged off. Too well-mannered to vocalize the disappointment visible in his face and drooping shoulders, he reminded me that under the rules of the Fair all entrants had to be present the night the winners were announced. Brushing him off, I truthfully assured him there was no chance of my winning.

At midnight our phone rang while I was still at the dining table writing to the DAR about freedom. I scrambled to find the phone before the ring tones woke my parents.

Mr. Shore was excited and stuttering, "Joan, you won first place! I called this late because I promised to bring you to Fresno in the morning so McClatchy Newspapers can take your picture with your project. They are going to pay for your trip to the National Science Fair in Flint, Michigan next month!"

The McClatchy Newspaper pictures show me standing in front of my project in my first Lanz dress, a black cotton print with a princess line outlined in oversized white rickrack. Mildred told me the line of the Lanz dress was correct for my height and that it photographed well.

The author in her first Lanz dress

A few weeks later, AFS announced that I had been selected to be an exchange student and was scheduled to leave home in June for Christchurch, New Zealand. When a reporter from *The Bakersfield Californian* called for a comment from my parents, my mother answered the phone. Not having any idea what the reporter was calling about, she told the reporter that she knew he was really an FBI agent trying to scare her. My mother had begun hiding our telephone again—sometimes in the refrigerator or the oven, locations she

---

* "Bakersfield Student Wins Top Prize at Science Fair," *The Bakersfield Californian*, April 16, 1958.

believed prevented the FBI from both trying to talk to her and listening to what she was thinking. Eventually the reporter reached my father at his research lab and he made a reparative statement.

Except for visits to Aunt Mary's house in Salt Lake City, I had never been away from home alone. My trip to the National Science Fair in Flint, Michigan was also my first airplane ride. My week's wardrobe was carefully packed into a new "World Traveler" Samsonite suitcase, half my height. Fully packed, I could hardly lift it. Inside were two new Lanz dresses, including a new one for the awards ceremony—a floral print in polished cotton, sleeveless and scooped low in back, where a five-inch-wide-apron-style belt was meant to be tied in a bow. Every edge was trimmed in white rickrack. It was a dress designed for a five-year-old. Mildred was away on vacation when I went shopping, hence my regressive wardrobe mistake.

National Science Fair contestants were housed in Flint's Berridge Hotel, where we were served three meals every day for a week. A survivor of The Great Fried Egg War of 1949, I ate everything on my plate at every meal, as if Commander Margaret had prepared the meals herself and was standing behind me with a sharp bayonet touching my spine as a reminder that I was obligated to clean my plate.

On the night of the awards banquet, my roommate struggled for a long time to button the back of my special Lanz dress. But it was impossible to solve the physics problem (collection of matter whose mass varies with time) created by my dutiful eating. I went to the banquet with my dress unbuttoned, wearing my roommate's dingy white cardigan to cover the gap. The sweater hung hump-like over the big bow as if I was harboring something illegal. I returned home with an honorable mention and ten extra pounds—persistent mementoes from that week of culinary satiety mixed with science.

The following month Mildred helped me select a suit on sale at Brock's made of beige ribbon taffeta for the DAR Awards Ceremony. Since I was going to "take tea" with sixty real ladies, Mildred suggested a straw pillbox hat and a pair of white kid gloves. She hemmed the dress while I went next door to Karl's Shoe Store for matching heels and handbag. I never told Mildred, but on the day of the tea, I began getting dressed by pulling on my Brock's flowered panty girdle. Incapable of returning anything, I didn't want the girdle to be a complete waste. That day in May 1958 was the first and only time in my life I put on every category of clothing I had grown up expecting to wear as a lady. The list was long: panties, girdle, nylon stockings, brassiere, slip, dress, high heels, hat, gloves, and handbag. In the pictures from that infamous afternoon, I look like a cross-dresser. That is, ninety-five percent resembled the style of Mamie Eisenhower and five percent hinted at Jackie Kennedy—statistical proportions the exact opposite of what I had hoped to achieve.

Not long after the DAR tea Mrs. Ryan called me into her office. She gently explained she had just received an urgent call from AFS officials. There was no family in Christchurch willing to host an American student of mixed religious background, at least not one as complex as mine. I had inadvertently slammed my own escape hatch shut by telling the truth.

However, AFS wanted to offer me a consolation prize. Mrs. Ryan explained that AFS was looking for a home for a female exchange student from Izmir, Turkey. The Nebraska family sponsoring her had reneged.

Mrs. Ryan spoke enthusiastically, "Joan Carol, I know how disappointed you must be, but I was wondering if having the Turkish student live with you might not provide some of the experiences you hoped to have in New Zealand!"

Devastated to learn AFS would not be helping me leave home, I fought back tears. Yet in a near fatal act of cowardice, I was unable to tell Mrs. Ryan that the consolation prize was not only unacceptable, but a dangerous poultice. I lacked the social courage to tell her that I had some very good reasons for wanting to leave home. Ironically, I no longer remember the circumstances under which my father agreed to the reversal or whether my mother was even consulted.

-23-

An Agent of the Turkish Police

In August 1959, my parents and I drove to Los Angeles. First we dropped "some important materials" at the offices of a certain well-known swimsuit maker (Rose Marie Reid) before proceeding to Los Angeles International Airport LAX to meet Ayse Tanner, our new housemate.

Holding a passport-like photograph of Ayse, I spotted her as she came down the steps of the plane and began walking toward me across the hot windy tarmac. Ayse had a stout figure and was my father's height—five feet, six inches. Her hair was dark and curly, her face covered with severe acne scars, inadequately disguised by a heavy coat of make-up. Ayse was wearing a suit and carrying a handbag that looked like they had been stolen from Queen Elizabeth's closet.

I have no idea what Ayse thought of us. My father had dressed in his best suit; my mother in her best dress. I was wearing a shirt dress with a Muslim-style head scarf as a defense against the brutal Santa Ana winds blowing that day. As soon as we tried to make ourselves comfortable in the overheated Fish-Car, I smelled regret—Ayse suffered from terrible halitosis and even worse body odor.

Ayse Taner's Arrival from Ismir, Turkey, August 1959

Ayse came from a country where she had been culturally conditioned from birth to feed my mother's madness. Already engaged to be married to a man my father's age, Ayse had been trained in Turkey to cater to every male whim, skills she immediately put to work.

Each night when my father returned home, Ayse would gently fling rose water on his face and hands, before escorting him to his recliner in front of the television. Next she would bring her host a glass of sweetened tea and the newspaper. While he read and sipped his tea, Ayse would massage my father's shoulders. She would then proceed to assault him with her most intimate gesture. Having previously filled my mother's soup pot with warm water, she would remove my father's shoes and socks in order to wash and dry his feet before anointing them with rose oil.

My father's mother had taught him to always be deferential to women, which made it almost impossible for him to keep Ayse at an appropriate distance. Also, he must have been suffering from a female attention deficit as big as my mothering deficit. Less than ten days after Ayse moved into our home, my mother became convinced that Ayse had been sent by the Turkish police to steal her husband. Further, my mother became convinced her daughter had signed on as a secret agent of the Turkish police. She was certain I had been part of the plot from the beginning; unfortunately, this was not far from the truth.

In an effort to keep Ayse at maximum distance from my mother, I arranged for her to be included in whatever I was doing, except work at Brock's. I tried very hard to find someone to entertain Ayse, anything to keep her out of the house. But if my father was expected home, she wanted to be there. Ayse's interest in boys or girls her own age was nil.

In late October, during Ayse's eighth week in our household, my mother developed a persistent fever, cough, and debilitating fatigue. Eventually she was diagnosed with San Joaquin Valley Fever, a fungal illness resembling mononucleosis and pneumonia. Nothing her doctor said persuaded my mother that her illness was anything other than attempts by Ayse and me to poison her drinking and bath water.

My mother's condition continued to worsen and by December she was hardly able to get out of bed. I finally mustered the courage to ask Mrs. Ryan for help.

My mouth as dry as the surrounding oil fields, I confessed, "My mother is quite ill and I have reluctantly concluded we need to find another placement for Ayse."

Mrs. Ryan was her usual genteel self. "I am already aware of your mother's situation, Joan Carol. Ayse is moving out at the end of the week."

Ayse had been to see Mrs. Ryan weeks earlier. I felt stripped naked, certain Ayse had told Mrs. Ryan and others that my mother was crazy. Ayse packed her bags and left our house on the day school adjourned for Christmas break, which was also the night of the EBHS Christmas Ball.

My escort to the dance was Dick Alton. I had stayed up until three in the morning for a full week before the dance sewing a formal dress from a ridiculously complex Vogue pattern on Nanie's Singer treadle sewing machine. My legs were so swollen from sitting and lack of sleep that on the night of the dance, I could hardly get my feet into my satin pumps.

When Dick drove me home, my heart was aching for Bart. Despite that longing, when I finally crawled into my bed, I realize that it was fortunate that Bart hadn't traveled by bus from Bozeman to Bakersfield again. I wouldn't have wanted for Bart to have seen his "Beautiful" the way she was the Christmas of 1959.

-24-

"The Perverse Mother"

By the time Ayse moved out, my mother's muscles were completely wasted. While the demons in her mind had grown ever stronger, her limbs had grown steadily weaker. My father finally asked her internist to make a house call. After his physical examination, her doctor expressed concern that perhaps the Valley Fever had invaded her brain. I overheard my mother telling him that she had been poisoned by her daughter, who was working for the Turkish police. I walked him out to his big white Cadillac and asked to see him at his office the following day. The next afternoon, as I sat across from her doctor, I could hardly form words trying to describe some of my mother's behavioral history.

But before I finished, he interrupted, "It sounds as if your mother has paranoid schizophrenia."

That was the first moment I learned the name of the disease destroying my mother's mind.

The week after Christmas, I told Mildred I needed time off for a research project, having already asked the reference librarian at Bakersfield Junior College to order copies of the most recent books and research articles on schizophrenia through interlibrary loan. In the deserted library, I was met by two bored librarians overly anxious to help.

After only a few hours of facilitated research, I was filled with a terrible fear that would define my life for another two decades—the fear that I would develop schizophrenia just like my mother.

As the symptoms unfolded before me, my mother provided haunting images for the medical narratives. Worse, every article by psychoanalytic theorists laid the blame for the development of my mother's schizophrenia at the feet of her mother, my beloved Nanie.

Dr. Frieda Fromm-Reichmann wrote that schizophrenia was caused by mothers who were "simultaneously over-involved with and rejecting of their children"—labeling those mothers "schizophrenogenic."[18]

Sitting at a long table in the nearly empty library, reading the literature under the surveillance of the two librarians, I felt as if the top of my skull had been blown open. There was wide-spread agreement among the experts that individuals with schizophrenia had been reared by a woman who suffered from a "perversion of maternal instinct."

Suddenly faced with a new reality, I realized, *"I am doomed! I am one of Blackie's kittens! It would have been better if I had died in my sack and never left the porch!"*

My ancient Aunt Mary hankie was soaked. One of the librarians noticed, and silently placed a box of Kleenex on my table.

Another psychiatrist, Dr. John Nathaniel Rosen, had joined Dr. Fromm-Reichmann in convincing his fellow psychiatrists that the mother of a schizophrenic had to be malevolent, even if there were no obvious signs of her evil nature. Dr. Rosen's theory was that the unloving mother completely explained paranoid thoughts. Apparently my mother was not imagining messages, but instead, she was remembering them from infancy. These experts were declaring my beloved Nanie was the cause of my mother's madness. If so, my life was without hope.

The threatening voices in my mother's head were, according to Dr. Rosen, simply recapping the messages from her mother, my Nanie, who, unable to love her, responded with feelings or behavior that amounted to: *"Be still! Be quiet! Be dead!"*[19]

Full comprehension was slowed by fear, but before I could reach the bathroom, a wave of nausea rose in my throat. On my knees, trying to clean up after myself with brown paper towels from the rest room dispenser, I felt the same as I had the morning I came upon Blackie's kittens on the front steps—uncontrollable revulsion.

Dr. Rosen was one of the few theorists I read that day or the next who proposed any kind of a cure. Since schizophrenia was caused by terrible mothering, he believed the cure was "good mothering." But where would my poor mother find a good mother to provide the nurturing she had never received? *Then I wondered where I was going to find such a mother. Perhaps Mildred would consider adopting me?*

According to Dr. Rosen, the therapist for a schizophrenic patient had to be "like a perfect mother" and it was necessary "to start over and raise the patient again." This hypothetical mother-therapist would have to become "the ever-giving, ever-protecting maternal figure" my mother had never known and there could be no shortcuts, no faking! The only way for my mother to recover from schizophrenia was for her to have a therapist capable of convincing her that her needs were understood and could be satisfied. Only then would my mother "dare to wake up from her ongoing nightmare."

I wondered, if she woke up, would she be able to start over with me? The average age of onset was about twenty-one for females, eighteen for males. Yet, none of the articles seemed to suggest any preventive treatments for the children of schizophrenics like myself. Why was that?

The next day I returned to the library for more of what had dominated my disturbing dreams and read that the key was convincing my mother she had a determined and insightful ally. Then, Dr. Rosen argued, my mother would "gain the sense that she is no longer alone, that she is understood." *I thought of Snowball teaching Blackie how to nurse her kittens.*

Leaving the library after the second day of devastating research, I made my way to a knoll on the edge of the Bakersfield bluffs overlooking the oilfields. Below me the Kern River was a trickle; above it I was drowning. The walls, labyrinths, and underground shelters I had so carefully constructed to protect me from my mother were dissolving into an indecipherable watery grave. Memories of my Nanie were of a warm, loving, humorous, and tender-hearted woman. *If my mother had grown up with Nanie, but had still become the mother I so dreaded, what was ahead for me?*

Since I had clearly been raised by a cold and paranoid mother, the experts said there was no way for me to avoid developing the same disease. Certainly, I must never marry or have children. If I were somehow able to survive my maternal inheritance, I had been convinced by these faceless expert doctors that it was my moral imperative to avoid passing it on.

Before I left the bluffs, I spoke out loud to the Heavenly Father I had left behind in Utah. I mumbled to myself, as if the onset of schizophrenia had already occurred, *"Dear God, please help me help Mom. And please, if you can, help me!"*

My mother was still too weak to prepare meals, so I heated two cans of Campbell's tomato soup for dinner, grilled three cheese sandwiches, and then carefully stirred a packet of Jell-O butterscotch pudding into a pan of warm milk.

At our usual places around the kitchen table, I found myself looking at my mother with fresh tear-filled eyes. I not only felt fear of her, but also, something like sympathy. Because of my library learning, I now knew the two of us were in the same leaky life boat. Neither of us had a paddle, a map, or a compass. Worse, there was no safe shore.

-25-

A Mother for My Mother

On New Year's Day, I filled our familial tub with rosemary-scented bath salts for my mother, trying to imitate a therapeutic gesture, as if I were Dr. John Nathaniel Rosen. After I helped my reluctant mother into the bathroom, she asked me to leave, firmly stating that she could manage on her own—her tone and fierce look made it clear she believed I was still working for the Turkish police, attempting to lure her into poisoned bath water.

After that contaminated transaction, I asked her internist to recommend a psychiatrist. I desperately wanted my mother to be cured and the only hope seemed to be Dr. Rosen's recommendation—"a mother for my mother." My first request, inaccurately translated by his receptionist, resulted in a referral to a physical therapist. Naively ignorant, I was happy when the receptionist gave me the name of a woman. Only after I skipped school one morning and crossed town on three different buses, did I discover the miscommunication.

The next day, I insisted on speaking directly to my mother's internist, who regretfully explained he didn't know of a psychiatrist to recommend. He suggested contacting the Kern County Medical Society or the Camarillo State Mental Hospital for help. Not only was there was no female psychiatrist in Kern County, there was only one male psychiatrist, and he specialized in forensic work. If my mother was going to be re-mothered, she might have to settle for a male willing to take a trans-gender part. Finally I asked the woman at the Medical Society if she knew how I could find a good psychiatrist.

"Look," she said, "if you are looking for a particular specialist, I suggest you call the UCLA Medical School for a recommendation."

Three calls later, someone at UCLA gave me the number for Dr. Norris H. Weinberg. There was a three week wait to see Dr. Weinberg. I was forced to tell the receptionist my mother would be coming for the appointment because, when she asked my name and age (then seventeen), she immediately announced that the doctor's policy required a parent to be present for the initial appointment. Trying to bring truth into my family required creative deception.

On appointment day, I skipped school again, taking a Greyhound bus to downtown Los Angeles, a cab to the doctor's office, and then faking my way past the watchdog receptionist.

"My mother is parking the car." I said. When she didn't look like she was going to buy this, I quickly invoked the name of my mother's internist: "He recommended I present my concerns directly to Dr. Weinberg without my mother."

That half-truth finally opened Dr. Weinberg's door. I expected him to look like Sigmund Freud—old, grey-haired, with a beard, holding a pipe. Instead Dr. Weinberg had no beard, no pipe, and looked to be about thirty. I talked too fast and long before pausing. I concluded by telling him about Dr. John Nathaniel Rosen's article, "The Perverse Mother."

"Since Dr. Rosen seems to be the only person who holds out hope for a cure, I was wondering if you would be willing to mother my mother?" a question that made me sound like a crazy woman.

Dr. Weinberg did not laugh or turn away. Instead he gave me two hours of understanding and permission to separate from my mother. He saved my life by giving me a prescription to live my own.

But first he told me, "Dr. Rosen's theories are popular at the moment, but they have not yet been proven."

Assuring me no one yet knew what caused schizophrenia, Dr. Weinberg said he believed the disease was caused by one or more environmental assaults triggering a genetic vulnerability, referring me to a 1928 study by Dr. Karl Menninger.[20]

Dr. Weinberg also reminded me that the role I was playing was really my father's responsibility, pointing out, although he needn't have, that I was a minor with no legal authority. While he believed my concerns about my mother were legitimate, he wanted to talk about how I could move forward with my own life.

"You may be at greater risk of mental illness, but whatever cruel and confusing messages you received, Joan Carol, you have become an impressive young woman. Clearly you are extraordinarily intelligent and brave to be trying so hard to save your mother. I want you to understand that the undamaged part of your mother, and clearly you see she has good parts, would want you to have the best, most satisfying life possible. That is the part of her you must take when you go away to college."

"But, Dr. Weinberg," I said, "while it is very comforting to hear you say no one knows what causes schizophrenia, I feel overwhelmed by the preponderance of medical and psychological literature condemning me to my mother's fate. I have spent most of my life struggling to keep a safe distance from my mother, yet I suddenly feel I am the only person who can save her! My father cannot seem to see how ill she is."

Dr. Weinberg leaned forward: "It is likely your father will continue in his denial, but no matter what happens, Joan Carol, you must remember that deep inside no mother wishes their child to give up on their dreams in order to save them. It is exactly the opposite. You are the one good thing your mother has created."

His words brought a flood of tears. Dr. Weinberg had Kleenex handy, moving the box closer to me.

As if he were channeling Elaine paying tribute to Blackie, he spoke with new emphasis, "I can see how hungry you are for your mother's love, Joan Carol. I assure you, she loves you in ways hidden by her disease. Don't ever forget that your mother wants you to be happy. To do this you must move on. She may even be better off with you out of the house. It sounds as if your adolescent development has complicated her reactions to the world. Try to remember this as you prepare to graduate. The best thing you can do for your mother is to create a life for yourself away from home."

As I left his office Dr. Weinberg handed me a book by Dr. Albert Schweitzer[21] telling me, "You seem to have a strong sense of caring for others. Why don't you read this? It may help you to imagine another way of life for yourself."

I felt lighter than air all the way back to Bakersfield. At the end of that bus ride, I realized I had a new role model. I too wanted to become a doctor and save lives, just like Dr. Norris H. Weinberg.

-26-

Changing the Distance

Ignoring my repeated pleas, neither my mother nor father went to see Dr. Weinberg. He had agreed to meet with both or either one of them in order to make an appropriate treatment referral. Nor did my father ever talk with me about the research articles I had given him, including Dr. Rosen's "The Perverse Mother." When I pushed my father to read what I had given him, he claimed he hadn't had time.

"Look, Daddy, I need to talk with you about what we can do for Mom and what I should do about my own life if these scientists are right," deliberately referencing "scientists" trying to trigger his interest.

"I don't think there is anything that can be done!" he said, abruptly leaving the room.

Conversations with my father about my mother always ended abruptly and left me in tears.

My mother's recovery from San Joaquin Valley Fever was extremely slow. But with Ayse out of the house and me on the launch pad, her more acute paranoid symptoms seemed to recede as she regained some of her physical strength. Another beneficial factor may have been the departure of the woman across the street.

When the "For Sale" sign went up, my mother said, "I knew her surveillance assignment would come to an end sooner or later."

Neither my father nor I responded. Our family communication patterns were always constrained by our futile attempts to smother my mother's paranoid thought processes.

Dr. Weinberg's words had temporarily saved my life, not my mother's. I could not wait to go away to college. I narrowed my choices to Radcliffe, Georgetown, or the University of California—either Berkeley or Santa Barbara. Money made my decision. UC Santa Barbara offered me a full-ride scholarship, including room and board. Radcliffe offered a token tuition reduction, while Georgetown came with a half-tuition grant. Wanting complete financial independence, even with a summer of full time work at Brock's, it would be a challenge to cover the costs at either Radcliffe or Georgetown. Instead of going east, I moved further west.

In September 1960, while my father watered the struggling gardenias, I packed my possessions into the Fish-Car. At the last minute, my mother announced she wanted to come with us. During the drive to Santa Barbara, I sat in the back seat stifling my excitement. As my father helped me unload the Fish-Car, I saw how much he had aged and my heart ached for him. Having shielded so many of my mother's worst behaviors over the past several years, I worried he would be overwhelmed as our upside down family structure was ruthlessly cut back. I felt as if I was abandoning him in the middle of an emotional desert without water.

No tears were shed during our awkward parting. Instead, covering our collective anxiety, we spoke about a plan for me to come home by bus for Thanksgiving. Then the Fish-Car disappeared down Mesa Road.

My new home was a single room in a surplus U.S. Army Quonset hut in Goleta, California. My room had its own sink and toilet; only the showers were communal. Still under the influence of Dr. Weinberg, I declared a double major in pre-med and Chinese studies because I had read Tom Dooley's *Deliver Us from Evil* and there was no African studies program. I modified the location of my plan to become a heroic doctor to Asia from Africa.

Feeling rich without tuition or other fees to pay, one Saturday I took the bus into downtown Santa Barbara and bought two more of those irresistible-only-to-me Lanz dresses—both had rickrack. I kept these two acquisitions a secret from Mildred, who otherwise guided me through the planning of my college wardrobe.

I found the course workload extremely easy to manage because I no longer had to splinter my time and energy between classes, homework, brides at Brock's, and extra-curricular activities, not to mention protecting both myself and the world from my mother, while denying I was a secret agent of the Turkish police.

The first course for my Chinese studies major was "The History of the Far East" taught by Professor Immanuel C.Y. Hsu. Professor Hsu's class was more like an interactive seminar—one in which I became an enthusiastic participant. Chemistry, calculus, composition, and German were also on my schedule.

Goleta's beach quickly became my favorite place to unwind. In mid-September, I came across an abandoned tabby female kitten with over-sized ears. She was clearly hungry, wet, and cold. I took her back to my room with every intention of finding a legal place for her to live, but by morning she had purred her way into my heart. I named her "Norrie" a poor disguise for my missing Nonny, but I spoke to my new feline as if she were my best friend. Norrie's presence in my room was a successful secret for six weeks. It seemed as if I didn't know how to live without having something to hide.

The inevitable moment came when Norrie escaped into the center hall at the same moment the Quonset monitor passed my door. Norrie was given an eviction reprieve until the end of the week. I couldn't imagine parting with her, so humbling myself, I called home.

Only after I begged, "Please, Daddy, please!" did my father reluctantly agree to take my rescued cat.

I found a ride to Bakersfield for Norrie with a student headed to Porterville for the weekend, tearfully placing her cardboard kennel on the driver's backseat, along with a box of supplies, (including Star-Kist tuna cans), and her favorite toys. Norrie made me look forward to going home for Thanksgiving.

Shortly after Norrie's departure, Professor Hsu, who was also my academic advisor, invited me to attend a violin concert in Santa Barbara. His invitation created an opportunity to wear one of the new outfits Mildred had assembled for me, still untouched since her selections were more suitable for Radcliffe or Georgetown.

My college closet was a reflection of Mildred's fashion mantra: "It is far better for a lady to be over-dressed, than under-dressed."

Perhaps it was the urge to wear Mildred's outfits that blinded me to any downside in accepting the invitation from Professor Hsu. After the first violin concert, we went to two symphony concerts, a play, and once to a French movie with sub-titles, followed by dinner at a French restaurant. It never occurred to me that my professor had a romantic interest in me; plus it was long before universities had begun to promulgate rules prohibiting socialization between faculty and students.

At Thanksgiving, I returned home to find Norrie had won my father's heart in the same way she had wooed me. Showing symptoms of early maternal depravation, Norrie had become obsessed with sucking on the soft fuzzy lining of my father's nylon windbreaker. She would crawl up his leg in order to snuggle inside and try to nurse on the lining. Images of Blackie and Snowball floated through my visual field the rest of Thanksgiving weekend.

My father and I prepared the turkey, gravy, and mashed potatoes. My mother made a Jell-O salad with canned cranberries, walnuts, and marshmallows. As was our tradition, except for the cranberries, there were no fruits or vegetables on the Lieberman table. Formal expressions of gratitude were also missing, but my father offered a standard prayer from his two-year stay in an Ogden orphanage, "Bless the meat, turn over your plate, and begin to eat!"[22]

After Thanksgiving, I returned to my studies and weekend cultural excursions with Professor Hsu. We met in his office to review my term paper the night before my organic chemistry final. Afterwards, Professor Hsu suggested we eat dinner together in the campus dining hall.

As we cleared our trays, he said: "How about a short walk on the beach?"

I agreed; I was sick of studying organic chemistry. We made our way down to the beach, stopped to take off our shoes and socks, and began walking south, each carrying our own footwear. Before any of the plentiful beach tar had begun to stick to the soles of our feet, Professor Hsu suddenly stopped.

Speaking in a calm even voice, Professor Hsu said, "I would like to ask your father's permission for your hand in marriage."

Professor Immanuel C. Y. Hsu, Santa Barbara, 1961

Had the wind and surf distorted his words? My mouth went dry, while my mind was racing. The professor had never even held my hand! Had he ever even touched my body? Perhaps he had taken my elbow to guide me towards our seats in a darkened theater. But regardless, the words he had just spoken meant there was absolutely, never, ever, at any time, going to be enough distance between us.

The only thing I managed to say was, "Please, Professor Hsu! You must not speak to my father! I have absolutely no interest in marriage!"

Shocked and discombobulated, I turned away from him and ran barefoot back to my room.

It was a miracle that I passed that organic chemistry final. I owed the fates further gratitude because it was my last final. After turning in my Blue Book, I called my father collect from a phone booth outside the library, telling him I needed to transfer to UC Berkeley immediately. His response was predictable. Such a transfer would not even be discussed until summertime, and, even then, he wanted me to know that he thought it was a very bad idea. After he refused further discussion, saying he was late for a meeting, I was forced to tell him about Professor Hsu's proposal.

Upon hearing the evidence behind my request, he quickly relented, "Okay, I'll be there by noon tomorrow."

The next day we haphazardly heaped my belongings into the Fish-Car. At my request, my father brought Norrie along for the ride. She lay on my lap all the way back to Bakersfield, nursing on the fuzzy nylon lining of my father's windbreaker, which I turned inside out and lay over my thighs. We listened in silence to a classical music station. I did not ask what, if anything, my father had told my mother. I did not want to know.

I worked in accessories at Brock's through the semester break. On my first day, Mildred and I met for lunch. Over tuna salad sandwiches, I told her the saga of Professor Hsu. Mildred's response was firm and emphatic.

"My dearest, Joan Carol. I have never understood why you have such an inaccurate picture of yourself. You have an extraordinary presence. I hope you will learn to protect yourself better in the future!"

I felt praised, ignorant, and at fault.

-27-

The University of California at Berkeley

In January, I left Bakersfield for Berkeley, forfeiting my full UC Santa Barbara scholarship—a financial loss raised by my father so often that if I had one dollar for each reminder, I quickly would have made up the difference. His repetitive reminders of regret heightened my determination to be financially independent.

Because UC Berkeley had no room for me in the freshman dorms, I was given an exemption. The only approved housing I could find was a room in a female-only boarding house on Durant Avenue, just off Telegraph Avenue. My roommate, a short, plump redhead, had grown up in Peru while her father was in the Foreign Service. I never determined whether she was unfriendly or sleepy. Whenever I came into our narrow room, her eyes were covered by a black satin sleep mask. She spoke only to ask me to be quiet. I learned to spend most of my free time at the library to avoid her hostility and to give her the space she so clearly needed.

The large size of UC Berkeley's student body made it hard to build social relationships, complicated by the gender imbalance and intense competition in the pre-med track. The first zoology lab set the tone. As I entered the lab room, two Asian-American males were having a tug-of-war over a black cat cadaver that closely resembled Blackie. The Asian-American students understood that a good cadaver meant the dissecting difference between "A" and "B" grades. The same hyper-competitive phenomena occurred in all the pre-med classes; many came to lectures equipped with cameras, snapping photographs of the slides as soon as the professors put them up on the screen.

My respite from competition was an art class. I was in the art studio drawing, lost in the light and sight of the Berkeley hills, when my stomach cramped at the sound of a familiar voice behind me.

"Joan Carol, I hope you don't mind me finding you here. I wanted to see if you were available to join me for dinner?"

Immanuel C. Y. Hsu had arrived for a surprise visit. Before escaping Goleta in my father's Fish-Car, I had written Professor Hsu a note and placed it in his faculty mail box. The draft written in my chemistry notebook read:

Dear Professor Hsu,

Please accept my apologies for running away last night. Your proposal surprised and shocked me. I responded like a silly school girl. I want you to know that I have never had a finer teacher or advisor. I deeply appreciate and respect your intelligence, your teaching skills, and your talent for and love of music. I hope that you will understand my necessary decision to transfer to another campus.

Having acquired a few more ounces of maturity in the six weeks since his surprise proposal, I suppressed an impulse to bolt from the art room. Instead I agreed to join Professor Hsu for dinner at Larry Blake's Restaurant on Telegraph Avenue. It was noisy and crowded, unlike the quiet I knew Professor Hsu preferred. I tried to be as kind as I could with my own truthful defenses as delineated by Dr. Rosen.

"It isn't you personally," I explained. "I decided a year ago that I would never marry for private medical reasons. I hope you can understand and will move on with your life. I want to assure you, I made the decision never to marry before we ever met!"

Professor Hsu's eyes filled with tears. He asked for the check before we had been offered a chance to order dessert, always my favorite part of dinner. We walked back to the boarding house in silence. Gently taking my hand at the door, Professor Hsu continued to maintain his formal and respectful demeanor, but gently kissed my cheek for the first and last time.

Professor Hsu's visit left me in a lonely and unsettled state, which writing in my diary did not abate. I was missing Nonny, who I hadn't seen since the winter holidays. On Friday night I took a bus across the Bay Bridge and the L car out to Taraval Street, where Nonny, then a freshman art student at San Francisco State College, told me to meet her at a friend's apartment. The gathering was crowded and chaotic; I left without having had any private time to talk with my best friend.

Riding the bus back to Berkeley, I looked out the window into the dark waters of San Francisco Bay. I caught a glimpse of my own denial. I was rapidly approaching the age when I could expect to develop schizophrenia, an outcome incongruent with my rigorous course of study. Too young to have a real bucket list, I began to make one. If I only had a few years or months left before the onset of schizophrenia, I didn't want to spend those years dissecting cat cadavers that reminded me of Blackie. Instead, I wanted to see more of the world before I lost my mind.

To move forward sometimes requires taking a step backwards. A week later I was working full time as a waitress in a bakery-coffee shop on Shattuck Avenue, a split-shift—from five o'clock until nine o'clock in the morning and from five o'clock until nine o'clock in the evening. I sandwiched my classes in between. To keep up with my courses, it was often necessary for me to study all night in the dining room of the boarding house, an arrangement perfectly suited to the needs of my narcoleptic roommate.

At the end of the semester, not only did I have a balance of five hundred dollars in my bank account, I had pre-paid my passage on the SS *Groote Beer*, sailing from New York to Rotterdam under the flag of the Holland America Line on September 9, 1961.

In June, I returned to Bakersfield for the summer to work at Brock's for Mildred Verhaag with the goal of doubling my travel money. I didn't tell my father about my European travel plans until after my grades arrived in the mail. In anticipation of his "No" vote, I came heavily armed: I was nineteen years old, had a high grade average, but most importantly, I had my own money. Whenever my father expressed displeasure with my planned trip and the interruption of my college education, I silenced him with a reminder of my schizophrenic-filled future.

By the end of June, Nonny and three other high school girlfriends from our EBHS gang* had decided to join me. They worked fiendishly saving money or coercing it out of relatives to acquire one thousand dollars, our amateur estimate of the cost for a three-month student-style tour. Group travel with known companions reduced some of the concerns felt by all our parents.

---

* Kim Brown, Susan Briggs, and Sherry Woods. See Biographical Index at www.OptimalDistance.com/Biographical-Index

Until the *SS Groote Beer* pulled up anchor in September, I filled my summer days with comforting counseling from Mildred; my nights with snuggling with Norrie and her first litter of four adorable kittens.

The author with Norrie with her kittens, Bakersfield, August 1961.

-28-

Europe and Africa

The photographs taken before we boarded the train to New York City show five young women dressed as ladies were then expected to dress when traveling on public accommodations. We wore our best dresses, girdles, nylons, and high heels. It took four days and nights to cross the country in a club car. We disembarked with our miserable support garments sealed to our bodies. Mildred Verhaag was right—girdles were the invention of a sick mind. My one and only Brock's flowered girdle was abandoned in our shared room in an un-starred hotel near Times Square, never to be replaced.

Dressed for Train Travel, Bakersfield to New York City, September 1961. Left to right: Kim Brown, Joan Carol Lieberman, Nonny Thomas, Sherry Woods, and Susan Briggs.

After serial showers, we walked ourselves over to Lindy's Restaurant, where we made ourselves sick by washing down big pieces of New York cheesecake with martinis. New York was then the only state in the country where eighteen-year-olds were legally permitted to drink hard alcohol. Naturally, we felt obligated to comply with this law upon arrival.

The *SS Groote Beer* sailed past the Statue of Liberty on September 9, her decks packed with college students from across America, including a group of Catholic boys from Spokane, Washington, with whom we had the West in common. Heineken beer was ten cents a bottle so when the *SS Groote Beer* docked in Rotterdam six days later, we were bosom buddies. Several of the Spokane boys were traveling because they too anticipated losing control of their lives—not to schizophrenia, but to Jesus, Mary, and the Jesuit priesthood. Most were testing whether their faith equaled the requisite celibacy—covering their sexual anxiety by joking about whether or not to shoot their "golden bullets." None of us had the courage to consummate anything so we made perfect mixed-gender traveling companions. Parting reluctantly in Rotterdam, we made a group pact to rendezvous in Paris in December.

The first travel task for our five-girl tour group was to lighten our loads. We had packed for Europe like we had dressed for Amtrak. We shed our heavy suitcases and four-fifths of their contents, storing them with Holland America in Rotterdam. We re-started our tour with reduced wardrobes and modified luggage (mine was a used World War II Army backpack).

The five of us toured Holland, Denmark, Norway, and Sweden together, relishing the economic freedom provided by our prepaid Eurailpasses and Europe's extensive network of Youth Hostels. According to Benjamin Disraeli, travel teaches tolerance, but group travel tested my patience. One or more of us always needed a restroom and we often behaved like the silly American schoolgirls. I likely was also regretting my biological destiny. None of my friends, not even Nonny, knew about my mother's mental illness. Nor did they have any idea that my desire to travel was based on my future fate as determined by Dr. John N. Rosen and his peers. A few weeks into our tour we had traded future goals in a Danish train car; mine was the only imagined future without a husband and children.

Nonny had fallen in love with Jim Ekedal, a San Francisco State classmate. She was totally preoccupied with writing Jim, talking about Jim, and looking for letters from Jim. We were in Munich's Central Train Station when Nonny's romantic preoccupations collided with my impatience. Our train to Salzburg was already on the platform, about to depart. Nonny, always a careful shopper, had been browsing the racks of postcards for "just the right one" to send to Jim, a task that apparently required the careful examination of every card.

Assuming the stern voice of a tour guide, I nudged, "Nonny, we need to board now or we are going to miss the train!"

Nonny exploded. "I am not done yet! We can catch the next train! Stop being so bossy!" she shouted.

Stung and stunned by Nonny's angry response, as the train to Salzburg began moving out of the station, I climbed aboard as planned, an impulsive decision permanently changing the distance between my four friends and me. I was now on my own.

In a strange coincidence several weeks later, I ran into my former travel companions in the gardens of *Schönbrunn Palace* in Vienna, Austria. Before my train pulled out of the Munich station, we had seemed as close as sisters. But as we awkwardly sat down for tea, the emotional distance between us had doubled.

Prior to our accidental encounter in Vienna, not only had I grown accustomed to traveling solo, I was savoring it. Being alone allowed me to drop my usual level of acute alertness, the one that assumed my survival depended on anticipating the behavior and meeting the needs of everyone around me. Instead, I kept company with my diary.

I had inadvertently gone into a bar for transvestites in West Berlin and crossed into the grey-walled desolation of East Berlin. Whenever there was no room at one of the hostels, I used my Eurailpass to ride through the night to another destination. I learned to ski at *Kitzbühel,* and, after the accidental encounter in Vienna, spent ten glorious days in a small pension in on the outskirts of Palma, Majorca. The outsider in me became fearless, perhaps the remnants of my early plan to run away from home.

Despite the estrangement, I made my way to Paris in early December according to the pact made on the *SS Groote Beer.* The five of us and our Spokane shipmates shared an assortment of rooms in the *Hotel Jeanne d' Arc,* then not even close to a one-star hotel, located at *34 Rue de Buci,* just off the *Boulevard Saint Germaine.*

Nonny left Paris soon after I arrived to spend December in Italy with her mother's relatives, but we reconciled before her departure. It turned out not to be the last time. I made monetary amends in Paris by buying her an extravagant dinner and a new dress at *Printemps* to wear in Italy, a form of appeasement first learned from watching my mother in an Ogden dress shop.

Kim Brown, the Spokane boys, and several other new travel friends decided to celebrate Christmas on Ibiza. Although I had already spent ten days on Majorca, I was savoring the laughter and companionship, so I hitchhiked to Barcelona with one of the Spokane boys. We caught a ferry to Ibiza on Christmas Eve where our group rented a small house for thirty-four dollars a month—my share was two dollars and seventy-five cents. The tiny rental house was located only a few yards from the beach, but was only used for cooking and washing our bodies.

Accompanied by the three guitar players among us, our voices rose and fell as we sang, discussed religion, philosophy, and politics until the wee hours before falling asleep on the soft sand, drunk on sangria and the sweetness of life.

-29-

Never Wanting to Go Home Again

Europe was a safe distance from my mother. I never wanted to go home again. Although I missed my father, I began imagining a short-term, modified version of the doctor dream inspired by Dr. Weinberg. Since the schizophrenic bomb inside me was likely to detonate any time, I thought I might go to Africa as a volunteer. The first American Peace Corps volunteers were being sent to far off places by President John F. Kennedy. American idealism was on the ascendency and I wanted to be a part of it while I was still sane.

My dream of volunteering as a medical assistant was received with enthusiasm by the *SS Groote Beer* group in the *Hotel Jeanne d'Arc,* but with distress by my parents. My father objected to the ongoing disruption of my college education; my mother apparently considered Africa to be a place with many poisonous snakes.

Kim Brown and Paul Swift (one of the Spokane boys, also known as Porky) expressed interest in joining me. We were introduced to Father Michael I. Gannon, an American Jesuit living in Paris. Father Gannon, then thirty-five years of age, took his sponsorship responsibilities very seriously. While Kim and Paul[23] soon changed their minds and moved on to other life adventures, Father Gannon found a position for me working in a medical clinic run by the White Sisters in Ouagadougou, the capital of what is now Burkina Faso, but was then the French colony, *Haute Volta.*

To obtain a visa allowing one to live and work in *Haute-Volta*, all applicants were required to pass the French government's language exam. My French was weak, so I needed a concentrated period of language acquisition. My wallet was even weaker so Father Gannon arranged room and board for me at a Catholic-run orphanage in the province of Dordogne. The orphanage, known as *Foyer de Bonté* (House of Kindness), was located at *Temniac* near *Sarlat-la-Canéda,* staffed by a nearly extinct order of elderly nuns. Vacationing priests were welcomed at *Foyer de Bonté* because the priests could celebrate mass in the

Temniac chapel, thus relieving the sisters of the necessity of going down the hill into *Sarlat* for mass every day.

Father Gannon, having been one of those visiting priests, asked the Mother Superior (known as *"Ma Jeanne Mere"*) if I could study there. She agreed if I would teach English to the children. My interactions with the nuns, children, and *Tanté Martha*, a ninety-two year old impoverished local aristocrat (who believed herself to be a descendant of Napoleon) allowed me to perfect my French language skills. As a *monetrice*, I taught English in the morning; *Tanté Martha* rigorously tutored me in French every afternoon; and my evenings were spent reading French medical texts.

Ma Jeanne Mere was a sparkling energetic leader of the *Foyer de Bonté* mother ship, but her crew was aging out. When she learned that I had no intention of marrying and having a family, *Ma Jeanne Mere* began praying for my conversion so I could marry Jesus Christ. She believed I would revitalize her order because she also erroneously imagined I would be followed by other American girls and much money. Extremely kind and charming, *Ma Jeanne Mere* took me to meet Josephine Baker[24] who lived nearby in the *Chateaux des Milandes* with her twelve adopted children. In contrast, my biological mother persuaded Mormon Church authorities to send two missionaries to *Foyer de Bonté* to persuade me not to go to Africa.

Ma Jeanne Mere called me into the library, speaking in the broken English she liked to practice on me, she said, "You have guests, my daughter!"

The two young Mormon Elders waiting to meet me wore the standard white short-sleeved shirts closed by black neck ties. Suddenly faint, I closed my eyes in a futile effort to erase their image. It was as if they were police from my past, who had come to arrest me for consorting with a house full of Gentiles.

As I stood in the entry looking at the two young missionaries, there was more than recognition of their shirts and ties. In a painful coincidence, one missionary had been my classmate at Gallatin County High School[*] in Bozeman. Facial recognition filled me with the same fearful sensation as when

---

[*] The missionary from Gallatin County High School was Roger Skinner; the second Mormon missionary was Robert Mahlen Dahl. See www.OptimalDistance.com/Biographical-Index

I had opened my Purple Satin Hanky Box to discover my father's hat was from Wickel's Men's Store. It was the type of coincidence that tested my agnostic confusion. The voice of childhood culpability reminded me that if there was a Heavenly Father, he was exceptionally spiteful.

The proselytizing pair had come to hand deliver a letter for me from President Rulon T. Hinckley, then heading the French Mormon Mission. President Hinckley wrote that, "Africa wasn't a very forward-looking country." He wanted to persuade me that I should "choose to expend [my] youthful energies among a more hopeful populace."

Both young missionary men also felt I should not spend my life energies in Africa, a country where black men were not allowed to hold the Mormon priesthood because of the color of their skin. After transmitting this priestly guidance, the pair asked *Ma Jeanne Mere* if they could show her and the other nuns a film strip used in their conversion tactics. Not only did *Ma Jeanne Mere* agree, she spontaneously invited them to spend the night and they happily accepted, even though I stood behind *Ma Jeanne Mere* signaling that they should decline. I felt a strong urge to strangle them.

After dinner, the sisters, all life-long celibates past sixty, sat attentively in the dining room as the two elders gave them their procreating spiel. I felt angry, debased, and powerless. My mother's long arm was trying to drag me back home using the Mormon Church just when I thought I was more than a safe distance away. I blamed her instrument more than I did her, knowing that contacting Mormon authorities for any reason was damning evidence of her fear and desperation.

Four months later, after bidding tearful goodbyes to *Ma Jeanne Mere* and the children of *Foyer de Bonté*, I returned to Paris to take the requisite French language exam, to be vaccinated for *la Fièvre Jaune*, and to obtain my visa. Once again, Father Gannon arranged rent-free lodging—a maid's room in the mansion of a Catholic widow on *Boulevard St. Germain.* My narrow fourth floor room had a rock hard bed, a small skylight, and an ancient shared bathroom down the hall. I spent most of my time studying French medical texts, including those about schizophrenia, dipping into Lawrence Durrell's *Alexander Quartet* for recreational relief.

---

* *Baroness Madame De Foucaucourt at 272 Boulevard St. Germain, Paris.*

Father Gannon, who was handling direct contacts with the White Sisters provided updates when he took me to dinner on several occasions. Having once wanted to be a jazz musician, Father Gannon also invited me to join him at live performances by the wondrous *Edith Piaf* and *ZiZi Jean Maire*. When I lacked funds for the expensive airfare to Africa, Father Gannon presented my case to the American Woman's Catholic Women's Organization of Paris, which agreed to cover the cost of my round trip airline ticket.

Nonny, Jim, and Paul were also in Paris that month. We spent time together, and one day I returned to my free garret and found a note from Nonny.

April 13, 1962

My Dear Miss Leebermansky,

I came to see you, but you weren't here. On the way I stopped and got some pastry – two cream-puff things and two long flaky things. When I found you were gone, my disappointment was so sharp, I promptly devoured two of the little beauties. They are stale. This upsets me.

So I sat on the brick floor under the skylight in front of your door and read my book for a while. My behind is very cold, almost asleep. Where are you? I need to discuss such vital things as spring fabric patterns, should high heels have little straps with buckles, or not? You know— earth-shaking stuff like that. Every time we get together, I always seem to be speaking of some goonky little problem of mine, and never listening to your Africa excitement and plans and yet, I really am listening with my heart.

Everyone says to you: "Oh, Joan, I admire you so much—it is such a marvelous thing you are doing, etc." Me, I stand there and say, "Gee, your coat is cute," or "Did you cut your hair?" and these things will change the destiny of man right? But listen closely to me, my beloved Leeb, and you'll know that I mean all the good things, really. I am truly happy for you and your satisfaction and joy is so evident. Please know that inside I am with you and happy because of your fulfillment.

I am living very honestly now and sex is not what I mean. It is good to find out that sex isn't the superb-extra-glorious thing everyone tries to make it seem. It is just part of life, and should be woven into the pattern as temperately as all the other parts are. Jim is very special, and we are beauty. My wise Leeb must realize this, though. Oh how different we are (you and me), and yet not so different at all. Maybe we can go out for lunch or window shop or try on raincoats at "Au Printemps" or "Galleries Lafayette" Ca va? Bon!

    Nonny

I had learned a few days before Nonny's note that she and Jim were having sex. I wanted to ask her more, but I felt extremely cautious around her, trying to avoid triggering a Munich-like response. The next day I took two freshly baked replacement pastries to the room Nonny shared with Jim and found myself looking at the two lovers with new eyes.

A week before my scheduled departure for Ouagadougou, in a *bistro* on *Rue Saint Dominique*, as I wolfed down an omelet, Father Gannon fell off his priestly pedestal while nursing a glass of *Pernod*.

Speaking with ardor, Father Gannon said, "Joan, I have fallen in love with you. I have begun preparations to leave my order. Tonight I want to persuade you to abandon your plans for Africa. Instead, I desperately hope you will come with me to Greece. I want very much to begin a new life with you."

My mouth opened reflexively, but no words came out. I stood up and began backing away from our tiny round table. As I did so, I bumped into the adjoining table of a smug Parisian man, spilling his just served, very hot *café au lait* directly onto his crotch.

He began screaming at me, *"Merde! Merde!* Along with some deplorable French labels for women.

Tongue frozen, I pointed toward the street exit. As I made my way across the room, every innocent moment spent with Father Gannon collected in my throat. As I pushed open the front door, drenched in shame, I glanced back across the room at my half-eaten omelet and met Father Gannon's eyes. What was wrong with me? How could I have been so blind, so ignorant? Yet again!

I felt filthy, as if covered in a dangerous unconscious seductiveness. Like the Biblical Eve, I assumed all the blame.

-30-

Practicing Medicine in Ouagadougou

Father Gannon and I had no further in-person contact. I delivered books and a small electric coffee maker he had lent me to his Jesuit residence at *42 rue de Grenelle*, leaving them at the front desk. He then sent me a petulant letter, scolding me for treating him so badly. In retrospect, I think he was trying to cover his sexual tracks, because it was followed by a letter of apology from his Jesuit supervisor offering to assist me in any way. It was another forty years before cases of sexual abuse and manipulation by priests became public.

While living at *Foyer de Bonté*, I had been vaccinated by a physician in *Sarlat* for Typhus, Typhoid-Paratyphoid, Plague, and Polio. I had to wait to be vaccinated for Yellow Fever (*la Fièvre Jaune*) in Paris because it was an attenuated live vaccine offered in only a few certified locations. I was vaccinated on April 13, 1962 at the *Institute Pasteur De Paris,* I have no recollection of any serious reaction.

On May 1, just past five o'clock in the afternoon, I boarded an Air France flight to Ouagadougou, psychologically bruised, but resolute. Early the next morning, as I stepped onto the tarmac, the temperature was over one hundred degrees. Like me, the White Sister waiting for me inside the terminal was petite and dressed in street clothing. She held a sign that read, "*Mademoiselle Leibermann.*"

After confirming my identity, she skeptically surveyed my backpack and tote, before leading me outside to a rusty Vespa scooter. With perfect French hostility, she instructed me to put on my backpack, stepped behind me, and, using a short rope, tied my tote to my backpack. Taking the driver's seat, she silently motioned that I should climb on the small jump seat behind her. The Vespa shot away from the curb so fast, I almost fell off the back.

A cacophony of sounds and smells rose from Ouagadougou's red clay streets, which were crowded with a wild assortment of animals and humans using a confusing mix of movement modes: human heads, babies tied to bodies, goats, wagons, decrepit Citrons, Land Rovers, hand-built bikes, scooters, and cripples crawling.

For the next seven weeks, I worked alongside Dr. Renee Rebout, a tall, French-born and trained doctor of middle-age. I slept in a closet-like room beneath netting in a small bed in her apartment.

On my third day of work, I realized Dr. Rebout was an alcoholic. *Le Soeurs Blanches* operated the *Chez Fany Bar* to support the clinic, providing a symbiotic relationship for the practice of medicine by an alcoholic physician. I learned a great deal and may have even saved a few lives, mostly by acting as a short stop for Dr. Rebout whenever her alcohol blood level made her more dangerous to patients than the ailments that had brought them to the clinic.

Chez Fany Bar, Ouagadougou, Haute Volta, 1962

White Sister behind the Chez Fany Bar, Ouagadougou, 1962

In the middle of one night, I accompanied Dr. Rebout on an emergency house call. We went to a clay and straw hut on the outskirts of Ouagadougou to assist a woman in labor distress. Dr. Rebout quickly determined the baby was in the breech position. The mother was screaming in pain, surrounded by her seven young children. After complicated charades, I persuaded the children to wait outside the hut in the dark.

In the end, Dr. Rebout attempted to save both mother and baby by performing an emergency cesarean under way-less-than-optimal conditions, including her too-high alcohol level. I learned more medical techniques during that eight-hour house call than most medical students do in their first internship year. Sadly, neither the mother nor her baby son survived.

The audience of seven children, who had silently crept back inside to witness the bloody death of their mother, fell into an unbearable chorus of wailing. The sun was up by the time we finished washing and wrapping the mother's body for burial. As we left the grieving children, part of my soul stayed behind with them.

Dr. Renee Rebout with patients, Ouagadougou, 1962

While I wept uncontrollably, Dr. Rebout was dry-eyed and definitive in stating her rule of life.

"In emergencies, it is not God, but human hesitation that determines who lives and who dies!" she declared.

In that humid bloody room, a mosquito carrying a powerful viral load, drawn by spilled blood, overcame whatever antibodies I had developed to Yellow Fever. Four days later I woke up feverous with a headache so severe I could hardly lift my head from my assigned pillow. I was desperately ill for a week before a blood test confirmed the diagnosis.

I had expected to be working with Dr. Rebout for a year, but soon was helped to board a return flight to Paris. Amid feverous goodbye hugs, leaving my books and clothing behind, I promised Dr. Rebout that I would return as soon as I recovered, but I never did.

While recovering in Paris, I learned Nonny had booked passage home on the SS Maasdam on June 28. Jim was going to continue travel on his own for another six months. Lonely, depressed, and disillusioned, Nonny's presence seemed the most readily available anecdote for everything that ailed me, so I booked passage home on the same sailing.

Upon being discharged from the American Hospital in Paris, I took the boat train to London to see Bart Penny, my high school sweetheart from Bozeman, now in the Air Force and stationed near London. We spent a night together in a London hotel. Although we did not consummate our relationship, we came close. Our reunion felt so easy, as if we had never been apart.*

---

* I still have two letters Bart wrote to me in 1962 while I was in Tucson; I think it is likely that I failed to respond. I heard nothing more from him until 1999 when Bart managed to locate me with the help of Janie Quarnstrom, an EBHS classmate. He called me from his home in Anchorage, Alaska and I learned that he had married a German girl while he was stationed in England and that they have two daughters. He has had a career in corrections and construction.

Bart Penny, U.S. Air Force, 1961

The author and Bart Penny, London, June 1962

I was scheduled to board the *SS Maasdam* when it docked at Southampton, one day after Nonny boarded it in Rotterdam on the other side of the English Channel. Without significant sleep during my reunion with Bart, I mistakenly took the wrong London tube and missed the boat train to Southampton.

Making my way to a Traveler's Aid station, I learned the only way I could catch up with the *SS Maasdam* would be to board a supply boat scheduled to service the ship off the coast of Cork, Ireland at five o'clock in the morning. Travelers' Aid somehow managed to get me a seat on the last flight to Dublin, even though flights were over-booked with horse racing fans traveling to Curragh for the Irish Sweepstakes Derby. They also reserved a rental car for me so I could drive to Cork.

Landing in Dublin at one o'clock in the morning, I discovered no one in Ireland under the age of twenty-one could rent a car. I finally persuaded a cab driver to drive me to Cork. When the cab pulled up at the quay in Cork, the sun was just beginning to peek over the horizon. Totally exhausted, I mistakenly left my backpack on the backseat floor of the cab, even though I had to reach into it to retrieve a hidden envelope containing the extra cash needed for my fare. In a zombie-like state, I boarded the supply tug seconds before it moved away from the dock, with only my fanny pack. Luckily it held my passport and remaining few dollars. All the forces of the universe were in agreement that I should never go home again.

Nonny seemed relieved to see me, but she was incredibly lovesick for Jim. Exhausted, depressed, and still symptomatic, I slept through most of the crossing, sometimes dreaming of my narcoleptic Berkeley roommate. I disembarked from the SS *Maasdam,* with less than ten dollars, still wearing the same clothes. Holland America had mistakenly failed to load my World Traveler Samsonite in Rotterdam. Nonny's mother met her dockside. They invited me to join them in Yonkers for that first night in America, but my mental state prevented me from believing the offer was genuine. Or maybe it was because I was my mother's daughter.

This is not the full story of my failure to meet my heroic quest to save lives in Africa, but I returned to America feeling a deep sense of failure, dreading everything and everyone waiting for me. I had lost faith in mankind. I should have been treated for clinical depression, but the only doctors I knew were on the other side of the Atlantic Ocean.

Letters from my father made it clear that my mother was now suffering from persistent paranoia. Six months after I left for Europe, my father had been forced to abandon his research in Bakersfield. He had asked for an emergency transfer because it was the only way he knew to reduce my mother's symptoms. The USDA transferred my father to Tucson, Arizona, placing him in charge of a project working on killer bees at the University of Arizona. My parents departed Bakersfield quickly. I was anxious to know what they had done with Norrie, but recognized it would have been childishly inappropriate to ask.

In Tucson my parents rented a furnished apartment before purchasing a new house made of dark red brick. Sadly, my mother began accusing my father of letting the FBI into the house to read her papers. It wasn't long before she believed my father's faculty colleagues were involved as well. The emergency transfer failed to produce even a short reprieve from her paranoia.

My father wrote that my mother had spent two days in a Tucson telephone booth believing it was the only safe place. She was also frequently driving across the Mexican border to Nogales, where she would spend time in various restaurants and public buildings, believing Nogales was a secure site because the Mexican police were too incompetent to tell the FBI where to find her. All of my mother's letters had been full of chilling threats mixed with scriptural warnings. When I finally boarded the SS Maasdam, there was another tightly typed aerogram from my father.

He wrote, "As soon as your mother heard you were returning, she has seemed to be more like her old self."

I was caught in a classic double bind. Dr. Weinberg said it would be better for my mother if I went away. My father said the prospect of my coming closer made her more normal. My stomach heaved as the physical distance shrank. At the Port of New York, after calling my father collect, I used my remaining resources to reach Central Park, illegally spending the night on a bench. In the morning, my father wired funds to an American Express on Fifth Avenue. I then boarded a train to Tucson at Penn Station, sans girdle, suitcase, or hope.

The heat of Tucson in August would not be much different than the heat of Ouagadougou, but in Tucson, my mother waited in a state of desperation. I had gone as far away as possible, only to learn it was going to be much harder than sawing off a gangrenous limb in Ouagadougou to sever myself from my mother.

-31-

My Mother's First Commitment

The train pulled into the Tucson station at three o'clock in the morning. My father was waiting on the platform in his usual stance, thumbs stuck between waistband and belt, still holding a thick ring of keys. Given the hour, his taciturn style, and my deep depression, our reunion was almost totally silent.

As he ushered me into the guest bedroom of their new home, he casually announced, "Dick is arriving today by bus at noon." Followed by, "I'm sorry, I intended to get out to our storage locker before you arrived. All your things are still there."

I couldn't fathom why, let alone how, an arrangement had been made for Dick Alton to welcome me home. Dick was the young man who forgot his wallet in Bakersfield on the night my mother tried to protect herself from surveillance by the woman across the street with cardboard window shields. When Dick had learned I was going to Africa for a year, he had written me a "Dear Joan" letter in Paris. Had Dick had reneged on his rejection? Later I learned that Dick had spent an evening in Tucson with my parents during spring training with the Cal Bears.

It was a fate-filled mistake that Dick ended up back in my life story, even though we had dated steadily during my junior and senior years of high school. An American archetype—tall, blond, blue-eyed, athletic, and an excellent student, Dick was the third-born of five children and a popular member of our EBHS social group.

Following graduation from EBHS, Dick had gone directly to Berkeley from Bakersfield, where he had been recruited to play baseball and football for the Cal Bears while studying engineering on an ROTC scholarship. In Berkeley, Dick and I had met for coffee a few times, but both of us were too busy for dating.

Dick's persona was the embodiment of neutrality. He participated in his Methodist Church youth group, but never spoke of his religious beliefs. He followed all the rules, but knew how to have fun. He was obedient and helpful to his parents, but not overly dependent on them. He was studious, but never bragged about his success. More than anything else in the world, Dick Alton enjoyed playing, watching, and talking about sports—especially baseball. His résumé was close to perfect, except I had never been in love with him, not even close.

What I most liked about Dick was his mother, Florence. She was a loving mother of five, who always made me feel welcome in her home. Dick's father Harry, the realtor, played a peripheral role, saying little, even when not absorbed in television. While I was in Europe and Africa, I rarely thought about Dick. He sent so many Hallmark cards to American Express mail stops, I developed a life-long dislike for this form of second-hand emotional expression. Not unlike my mother, I felt as if I was being followed by something I was trying to leave behind.

In the guest bedroom of my parents' new home, I struggled to fall asleep on one of two unfathomably uncomfortable day beds. Four hours after my arrival, I opened my eyes in fright in the middle of a terrible nightmare. Coming to consciousness, I realized I wasn't having a nightmare. My mother was standing over me with a big butcher knife in her right hand and a murderous look on her face.

I was so frightened I wet the bed for the first time in twelve years, screaming, "Mom, what are you doing!? Give me the knife!"

Without saying a word, my mother's face showed a psychological shift into another reality. She silently handed me the knife before walking out of the room.

Shaking with terror, I took the knife into the bathroom and locked the door. While taking the longest shower of my life, I woke up enough to recognize my mother needed to be hospitalized. Her symptoms were far too acute and dangerous to be ignored.

To maintain adequate courage I silently repeated Dr. Rebout's mantra over and over: "In emergencies, it is not God, but human hesitation that determines who lives and who dies."

Desperately in need of a change of clothes, I had to improvise. Following my long locked-door shower, I took my father's robe from the back of the bathroom door and went into the kitchen. Utterly unable to imagine selecting anything from my mother's closet, I asked my father's permission to borrow a few items from his. After I dressed without undergarments in one of his long sleeved

Arrow shirts, and rolled up the cuffs on a pair of his washable khaki slacks, I felt more human.

Breakfast at our old Formica table was strained by my mother's murderous behavior. I was starving, but feared her special pancakes. My father had made Cream of Wheat, so using him as my taster, I ate what he left in the pan. Seeing my choice, my mother went into their bedroom, slamming the door behind her. After her abrupt exit, my father and I sat in the living room, awkwardly working at becoming reacquainted. After less than thirty minutes, he suggested I drive him to his laboratory so I could use the Fish-Car to pick up Dick.

My reunion with Dick felt like yet another burden. Dick and I went to a restaurant because he was hungry. I ordered what he ordered and ate it all.

Brutally blunt, I told him, "Please, go back to Bakersfield tomorrow; I can't be in any kind of relationship now."

I also told Dick for the first time what was wrong with my mother and what I intended to try to do about it. That is how Dick Alton came to be the ambulance driver for my pseudo-medical practice in America. That afternoon, while Dick took a long nap on the dry, unstained daybed, I drove to the Tucson Medical Center. Still disguised in my father's wardrobe, I made arrangements for my mother to be admitted the following morning. I was only able to do this because it was a few years before well-intentioned civil liberties lawyers made it impossible to commit a family member on anything more than a twenty-four-hour hold.

On my way home, I stopped at Levy's Department Store to make a few critical additions to my temporary wardrobe, particularly underwear.

As the sun began to set, the four of us went to a nearby restaurant for an extremely awkward dinner—the only time I ever felt grateful for baseball talk. Afterward, in mixed emotional states, we all went to bed early. Dick and I slept on separate daybeds in the same room, where I had also hidden all the knives and scissors I could find in the Levy's shopping bag beneath my daybed.

Using deliberate deception the following morning, I asked my mother to show Dick and me a few Tucson landmarks. But first we drove my father to his lab, where Dick and I were introduced to his colleagues, while my mother remained in the car. When we returned to the car, I suggested Dick drive. I distracted my mother with pseudo-questions about Tucson, while Dick drove directly to the entrance of the Tucson Medical Center following directions I had previously given him.

My mother tried to run away as soon as two male attendants approached us their hospital greens. As she opened the Fish-Car door to escape, the attendants each took her by an arm and started walking her toward the entrance.

Turning her head back toward me, she started screaming at the top of her lungs, "You are still working for them! I knew it! You've been helping them spy on me all this time, haven't you! God's angels will bring their wrath down upon you with all His might, young lady! You will never again have a moment of peace! You know what they are after don't you? If they find out that the poison snakes are still in you, they will kill you. I've been trying to protect you, but now you have destroyed everything!"

Her voice was so ferocious people walking in the vicinity froze in their tracks. Prior to Tucson, Dick had only deceptive glimpses of Margaret Lieberman. The previous day, when I told Dick my mother was suffering from paranoid schizophrenia, his face had exemplified disbelief, as if I was the crazy one, not my mother. Now his face was ashen.

Dick waited an hour while I finished the admissions paperwork, my first exposure to the value of my father's Federal health insurance benefits. Dick and I rode in silence to the Tucson bus station where he purchased a return ticket for a bus departing in ninety minutes. We walked to a nearby Mexican restaurant for the interim. It was easy to promise Dick I would call him when I was back in Berkeley, but harder to feign interest in the current team capacities of the Cal Bears or his course descriptions. I could hardly wait for him to be gone.

When Dick reached out to embrace me before he boarded his bus, I said, "Please find someone else!"

This was not a generous gesture on my part; I felt absolutely nothing.

I waited until the end of my father's workday to tell him what I had done with his wife and why.

His subdued response was limited, "I was looking for a knife this morning, but couldn't find one."

Never known for overreaction, my father remained mute through the remainder of our dinner time in a cafeteria he and my mother frequented offering five flavors of Jell-O.

As we climbed back into the Fish-Car he said, "Whatever you think is best. You're the doctor in the family."

By the time we returned to the house, my mother had been given her first dose of chlorpromazine or Thorazine.[25] We weren't allowed to visit her for forty-eight hours. Two days later, when we cautiously entered her room, she greeted us with what I judged to be extremely good news.

"It is so quiet here where they can't talk to me! Where do you think they have gone?" she asked in a tone of authentic curiosity.

While my mother's voice sounded confused—her facial expression was mystified, not menacing.[26]

My mother remained in Tucson Medical Center for twenty-seven days; she was discharged four days before I left Tucson for Berkeley. She swore, as so many schizophrenics do after their first treatment with an antipsychotic drug, that she would take her medications because she now understood how important doing so was to her survival, as well as to that of her husband and daughter. Scheduled to see her psychiatrist weekly, my mother even told us she thought she could trust him.

Just before I boarded my plane to San Francisco, she moved toward me and gave me an extremely rare hug. Despite that moment of medicated maternal closeness, as soon as the plane took off, the growing distance from my mother gave me a great sense of relief.

-32-

Thanksgiving in Berkeley's Tilden Park

Sailing home from Europe, I felt as if I was moving back into a futureless limbo utterly lacking the distractions provided by exploring Europe and heroic helping in Africa. I decided to go back to college because my father expected me to and because Berkeley provided some of the distance I needed from my mother.

At Berkeley I faced increased financial challenges. My parents' move to Tucson meant I was now an out-of-state student. My father tried to persuade me to transfer to the University of Arizona in order to avoid the increased out-of-state tuition, but relented as soon as he learned that the University of Arizona required parents to have been taxpaying residents for a full year before student dependents could claim in-state status. The difference in tuition costs was negligible.

My plane landed in San Francisco before noon. By nightfall I had rented a furnished room in Berkeley in the modest home of Mrs. Lindstrom, a widow. My second floor room on Derby Street at the intersection with Warring Street had a north facing window with a view of Tilden Park. There was a single bathroom down the hall shared with two other roomers—both males. There were no kitchen or telephone privileges, but Mrs. Lindstrom said she would take emergency messages and leave them on the front hall table.

On my second day in Berkeley, I applied for a position as the nighttime supervisor at the Doe General Library. The advertised work shift was from five o'clock in the afternoon until midnight. After checking my references, including Mildred Verhaag, I was hired and put in charge of a forty-person staff, most of whom were graduate students.

It was not until I took a French placement exam, earning sixteen foreign language credits, that my father conceded my period of travel had some value. Those credits meant I was less than a semester behind in my course-work. During that same telephone conversation, we made a mutual decision that it was too expensive for me to come home for Thanksgiving.

Three months passed. I had not seen or spoken to Dick since his bus pulled out of the Tucson station. Just before Thanksgiving, I saw him walking across Sproul Plaza. I quickly changed direction, but Dick ran after me. As we stood facing each other, disrupting the flow of human traffic like two rocks in a stream, Dick asked if I was going home for Thanksgiving because he wasn't.

"Let's have a meal together so that we have time to catch up!" Dick said, as I imagined another overly-detailed sports report.

On a sunny Thanksgiving afternoon, Dick and I shared a container of fried chicken and two beers in Tilden Park—both the menu and location were within our separate limited budgets. As we lay on our backs on a bed of dried Eucalyptus leaves watching crisscrossing jets and the birds of Tilden Park, Dick began kissing me.

I don't know why I didn't say, "Please stop, Dick." Blaming biology or my perpetual starvation for physical contact seems an inadequate excuse. Nevertheless, by the time the fog could be seen approaching the Golden Gate Bridge, there were dried eucalyptus leaves in my hair. We had "gone further" than either of us had before—past second base, but I believed whatever had happened between us physically was not sex. For me, it bore no resemblance to the imaginary sensations promoted in movies or books.

A month later, when I flew home for Christmas, my mother was still taking her Thorazine and my father seemed more engaged with killer bees. My mother's only Christmas tradition was to make chocolate rum balls. She brought a plate of them to the coffee table while we all watched the Ed Sullivan Show. Even the smell nauseated me. I often felt queasy in the presence of my mother, so I ignored my body.

-33-

"She Has a Turkey in the Oven"

I assumed my Christmas nausea was another after effect of Yellow Fever. Back in Berkeley, the nausea and fatigue made me feel as if I was still in Ouagadougou. After I vomited into the gutter on Warring Street, I went to the student health services. Because blood tests revealed abnormalities in my liver function, I was referred to the UC Medical Center in San Francisco.

The Chief Resident, Dr. Maurice Fox, conducted my examination in the presence of ten residents and interns, who were told to palpate my nodes for themselves. After taking an extensive medical history, including detailed questions about my work in Africa and the treatment of my mother for paranoid schizophrenia with Thorazine, Dr. Fox rolled his stool close to the exam table.

"Ms. Lieberman, I am sorry to tell you this," Dr. Fox said, "but I think it is likely you have Hodgkin's lymphoma."

He then wrote orders for additional tests, including a biopsy of several nodes. I spent six festering weeks believing Dr. Fox's preliminary diagnosis, returning twice for more tests and nodal biopsies.

During my third appointment Dr. Fox, said, "Ms. Lieberman, the good news is you do not have Hodgkin's, although you are still suffering the after effects of Yellow Fever. However, your test results show you are pregnant."

How was that possible? How could I have had sex without knowing I was having sex! I couldn't fathom how the awkward fumbling between Dick and myself in Tilden Park constituted sexual intercourse. From the sticky wetness on my underpants and pubic hairs, I knew Dick had ejaculated, but I had no memory of penetration. Neither had I noticed any blood. Was it possible that Dick's sperm had crawled up my cervix on their own?

My mother's paranoid projections and Dr. John N. Rosen's theories were intermixed with all the old morality lessons encoded in Mormon teachings. For the first fourteen years of my life, I had been indoctrinated with beliefs about the importance of chastity. The texts of LDS pamphlets came bubbling to the top of my psyche: *Be the Girl of Your Dreams, Someday You Will Marry*[27], and *The Symbolism of the Rose.*[28]

More critically, even though I was not the first unmarried woman to become pregnant, I knew of no one whose future was as blunted or as dangerous as mine. A tsunami of sexual shame washed over my psyche in a destructive moral aftershock. Raised and still living in an era when a woman's virginity was her most valuable asset, I believed that losing it made a woman worthless. Not only did I feel worthless, I felt dangerous. Dr. John N. Rosen was winning his argument with Dr. Norris H. Weinberg. Still unable to find published research pardoning me from my maternal inheritances, I felt my only option was suicide.

I kept Dr. Fox's premature diagnosis of Hodgkin's disease as a rationale for my intended suicide in an effort to protect myself, as well as Dick, believing that cancer was a less shameful and more understandable motive for suicide than an illegitimate pregnancy or the predicted schizophrenia. Perpetuating the initial Hodgkin's diagnosis, I lied to Nonny and other friends about my condition and lost their trust, as well as their respect.

What I did not know was that my mother had known I was pregnant the minute she saw my face at Christmas. On the first night I was home in Tucson, as soon as my parents were alone in their bedroom, my mother made an emphatic whispered announcement to my father.

"Frank, she has a turkey in the oven!"

-34-

Making and Breaking Vows

It had been three years since I learned about schizophrenia in the library of Bakersfield Junior College. Every thought, feeling, and dream I had passed through my self-designed schizophrenic detection system. The device was now so thoroughly hardwired into my nervous system, I was in a perpetual state of hypervigilance. After Dr. Fox told me I was pregnant, the news instantly resurrected the vows I had made on Bakersfield's bluffs to some spiritual shadow. First, I would never become a mother, and second, at the first sign of schizophrenia, I would interrupt the disastrous pattern by ending my life.

After my initial shock and panic, I began to feel strangely relieved. Conception meant the end of both denial and pretense. I had been living on borrowed time, working extraordinarily hard, behaving as if I had a future. Now I was free to tidy up my life, and to choose a moment and method of suicide that was right for me.

From my first encounter with Dr. Maurice Fox, I had found him to be extraordinarily appealing. He was kind, gracious, and had a lively sense of humor. He had expressed genuine interest in my African experiences. I had watched him carefully through four appointments, as he skillfully related to me as both a patient and a human being, while simultaneously being an effective teacher for a gaggle of nervous medical students.

Nor had Dr. Fox ignored my visible shock. He had taken time to assure me that the baby would be fine, despite the remaining dysfunction in my body. He had also made a personal referral to his favorite OBGYN resident. Knowing I was not married, he did not ask about the father of the fetus. What he didn't know was that I had no intention of becoming a schizophrenic mother.

Suicidal intentions are nothing if not preoccupying. Three weeks passed in which my ability to concentrate, sleep, or interact authentically with others fell to all-time lows. Dr. Fox was the only person in the world who knew both of my debilitating secrets. I believe this is what motivated me to ask for his help in obtaining a prescription for drugs that would end my life without pain and suffering for the fetus or undue trauma for my father.

Having asked for a private consultation with Dr. Fox, I was given his last appointment on Friday, March 1, 1963. Dr. Fox was behind schedule so it was almost six o'clock when he finally ushered me into his small office.

"Miss Lieberman, it is good to see you again! How are you feeling?" he asked.

"Physically better, psychologically worse," I responded. "As you may remember from taking my history, my mother suffers from paranoid schizophrenia. Having read all the current research, according to the collective expertise of the medical community, I will soon suffer the same fate. In 1959, when I first learned about schizophrenia, I vowed never to marry or to have children, not wanting to live like my mother or to perpetuate the disease. I have spent the past several weeks trying to tie up the loose ends of my life in preparation for suicide."

Dr. Fox stopped writing his patient notes as soon as I said the word "suicide."

"My initial plan, admittedly made while I was in a state of shock," I explained, "was to jump off the Campanile.[29] But I don't want the fetus to suffer or for my father to have to deal with a messy public aftermath. So the reason I am here today is to ask you to help me find the right pharmaceuticals for a less dramatic demise. Could you please help me acquire medication for something that would produce a painless death for the fetus and a more private one for my parents? Please, I don't know who else to ask!"

Dr. Fox was calm. "What about the father of your child?"

"Well," I responded, "he was a high school classmate of mine and is also a student at Berkeley. He has no idea I am pregnant and, in fact, may still consider himself to be a virgin, as I did until you told me otherwise. He and I engaged in nostalgic heavy petting in Tilden Park on Thanksgiving Day. It certainly never occurred to me that we had engaged in sexual intercourse, and I would be very surprised if he labeled our groping as sex. I have never been in love with him and have no intention of marrying him, but neither do I want to ruin his reputation, his academic status, or his athletic career with this news."

After speaking at length about my options and the illegality of lethal prescriptions, Dr. Fox said, "Joan, I've been on service for twenty-four hours. I am tired and starving for a decent meal. I should call the Psych Unit and have you admitted for an evaluation, but I think it would be more rational and helpful to both of us, if you would join me for dinner so we can continue this conversation."

I followed Dr. Fox outside where he hailed a cab to *Fior d'Italia* in North Beach, which he said was his favorite restaurant. The head waiter greeted him as if he ate there every night. Dr. Fox ordered two entrees, while I, a student short on funds, stifled my appetite, and instead imbibed his tender, therapeutic persona.

During dessert Dr. Fox made a critically important argument against the theories of Dr. Rosen and other psychiatrists who believed schizophrenia was caused by the type of mothering one received.

"Look, Joan," he said, "if inept or cold mothering caused schizophrenia, the majority of people in America would be locked up in mental hospitals, including myself!" He laughed before continuing. "You mustn't sacrifice your life or that of your child in expectation of succumbing to schizophrenia. No one has any scientific proof of such a crazy correlation!" He spoke with compassionate conviction.

The hormones of pregnancy, mixed with Dr. Fox's insightful perspective, triggered weeping. The attentive head waiter kindly brought me a small hot towel, as if directed to do so by Aunt Mary. It was after ten o'clock when Dr. Fox paid the bill, refusing my handful of proffered singles. There was a chilling drizzle as we began walking toward my bus stop. Tears were still rolling down my cheeks when Dr. Fox stopped, took me in his arms, and kissed me. What I believe may have technically constituted a swoon passed through my body.

We were near Washington Square Park when Dr. Fox hailed a taxi, telling the driver, "Palace Hotel, please!"

Inside the taxi, all Dr. Fox said was, "If you will, I want you to spend tonight with me. I want to try to remind you why life is worth living, no matter what the future holds!"

It was not hard to put myself under Dr. Fox's care. At that moment, had he asked, I would have followed him unarmed into a war zone. Although shamefully underdressed for both *Fior d'Italia* and the Palace Hotel, my mental state put any concerns about my amateurish rickrack style in the waste can. Keeping a discrete distance from Dr. Fox as he stood at the front desk registering for a room, I surveyed the grand interiors of the Palace Hotel, feeling not one-ounce of Utah embarrassment. I also deflected the leer from the bellboy who escorted us, luggage-free, to our room.

After Dr. Fox made love to me, I felt transformed. I also understood why I had not believed Dick and I had actually had sexual intercourse in Tilden Park. After Dr. Fox's initiation, I enrolled in his therapeutic weekend workshop for extra credit in human anatomy and psychology, astounded by both his expertise and teaching methodologies. I didn't go back to Berkeley until Monday morning.

After my sexual awakening, my mind resembled a kaleidoscope—in constant motion between having Hodgkin's, to being pregnant, to wishing I was pregnant with Dr. Fox's baby, to methods of suicide. Time was both frozen and slipping out of my hands. The weekend of marathon sex temporarily paralyzed my desire to die. My brain seemed to have swapped places with my uterus and was suddenly ravenous for protein, particularly tuna fish. Procrastination, muddled thinking, and wild deceit dominated my behavior. Whenever I tried to focus on an appropriate method of death, I found myself wishing for another sexual marathon with Dr. Fox.

Perhaps I was not alone in my longing for a repeat performance or maybe Dr. Fox was simply following up with a suicidal patient. Over the next three weeks he left so many telephone messages with Mrs. Lindstrom, she finally told me I had to put a halt to the calls. Then on the afternoon of March 21, 1963, I found two letters from Dr. Fox asking to meet again.

That night, near closing time, as I sat at the library turnstile trying to draft a response, someone approached. Looking up in expectation of checking out a book for a student, I instead saw my mother. Not surprisingly, I wet myself again.

-35-

My Mother Saves Her Progeny

I asked my mother to lend me her raincoat so I could make a semi-dignified exit. She followed me past the night guard out the front door to the grand entry steps, where we both sat down.

My mother immediately announced that she had already made contact with both Dick and his parents. I had not seen, nor spoken to Dick since our Thanksgiving Day picnic. My mother still seemed to be on Thorazine, but she had a very different presence, one with a strong military flavor.

"I have told Dick of your pregnancy!" she said. "He has agreed to marry you in order to give the child a legitimate birthright!"

My face flushed as I tried to imagine Dick's demeanor when Margaret Lieberman showed up at his door.

"Why did you assume that Dick was the father?" I probed.

"I called him in January to ask if he had ever had sex with you and he told me the truth!" she replied.

Dick had not only spilled his sperm, but spilled the secret! I felt furious! He wasn't ignorant of my mother's condition! How could he have been so stupid? By not warning me, his behavior was a double-play betrayal!

Struggling to remain calm, I spoke slowly, "I am not going to marry Dick."

I did not explain that I was planning to kill myself and my unborn child. Nor did I tell my mother that a brilliant and charming internist was doing everything in his power to dissuade me. Despite my hormonal fog, I still had some instincts about how to handle my mother. It was my father who brought tears to my eyes.

"Does Daddy know?" I asked.

"Yes, he has known since Christmas," my mother responded.

How was that possible? No one had known at Christmas! Then I remembered the power and range of my mother's antennae. The psychological boundaries of those suffering from schizophrenia are so thin and porous, their paranoid worries often contain invisible fragments of reality. For this reason, many therapists find it helpful to have a schizophrenic patient in group therapy as a kind of stealth barometer.

My mother's fierce voice continued, "You have an absolute moral obligation to your child to give it a proper name! What you do after that is up to you. I want you to understand that I will not allow my grandchild to be born illegitimately! It was you who made this terrible moral mistake and now you must do everything in your power to rectify an unforgivable lapse of judgment on your part!"

The tone and syntax of my mother was that of a stern Mormon elder.

It was very late that night when I spoke to my father on the telephone. He had just driven into Bakersfield where he was spending the night at the home of Florence and Harry Alton.

"I think it is best for you to do as your mother says," he intoned, so softly I could barely hear him.

I could sense his disappointment flowing toward me where, somewhere in the middle of the San Joaquin Valley, it met the river of shame spilling out of me. Then my father made the most idiotic statement I ever heard emerge from his mouth.

"When you went away to college the first time, I meant to tell you . . ." he paused, then after a long sigh, continued "to keep your panties up, your dresses down, and your legs crossed at all times!"

I hung up without saying goodbye, collapsing in a mixed-fit of hysterical laughter and sobbing on the gummy floor of the phone booth in front of my mother's Berkeley motel.

The next morning I met Commander Margaret at my former place of work, the bakery/coffee shop on Shattuck Avenue. Over a bear claw and hot chocolate, she proceeded to detail her orders for my marital deployment. UC Berkeley would not expel me because of my pregnancy. For some reason, she felt compelled to point out that I would have been expelled from both Radcliffe College and Georgetown University for this kind of moral infraction. It was a decade before *Roe v. Wade*. Abortion was illegal—a criminal act.

My only option was to keep my cool and execute my terminal plan sooner, rather than later. But I only had about twenty-four hours before my mother intended that I walk down the aisle of St. John's Presbyterian Church located a convenient two blocks west of my rented room.

My mother had done her license homework, explaining, "You and Dick can get your blood tests after the ceremony so long as the signatures on the marriage license are not witnessed until then. I want to take you to San Francisco today for a dress at Macy's and a few other things you will need. Dick's parents and your father are bringing some furniture in U-Haul trailers for the apartment."

"What apartment?" I asked, holding my own hands in a futile effort to stop them from shaking.

"I rented a studio for the two of you on Haste Street the first day I arrived," she said.

I had never seen this kind of behavior in my mother. It made me wonder if a new drug had been added to her treatment regime. She resembled a distorted, mutated version of Mildred Verhaag, intermixed with a heavy dose of Commander Margaret, last seen in the Great Fried Egg War of 1949.

My character has many flaws—the most consequential is a perpetual punishing culpability for any ill. Some guilt was imbued by my mother's belief that my body was filled with poisonous suffocating snakes and some from my having masqueraded through Mormonism in order to avoid being murdered by her. My sense of self was destined to become deformed and distorted.

I decided that if I pretended to obey Commander Margaret's orders, I could die as soon as the two pairs of parents left town. Dr. Fox was becoming shrouded in a fantastical fog. Dick not only would understand, he would be relieved! The proof of our powerlessness as prospective parents was that Dick and I still hadn't even spoken by telephone.

I did not accompany my mother to Macy's for a wedding dress. I had two exams that day and was underprepared for both. As we sat across from each other at a table I had previously wiped clean many times, I kept imagining myself running out the door, straight up Bancroft Way, to Barrow Lane, over to South Drive, where despite the public messiness, I would jump off the eighth floor observatory of the Campanile!

This solution alternated with another vision where I saw myself running out the door to catch the San Francisco bus, one of which was idling in my line of site on Shattuck Avenue at that very moment. The latter scenario ended in the arms of Dr. Fox. These alternating escape plans confirmed my worry that I was on the verge of schizophrenia.

My father had driven from Tucson to Bakersfield in the Fish-Car, pulling a small U-Haul filled with household furnishings. He and Dick's parents were now caravanning their way to Berkeley. That afternoon, while I was being unfairly tested in biochemistry and advanced calculus, the parents of the groom and the father of the bride arrived at the Haste Street studio apartment. Joined by Commander Margaret, the foursome unloaded an assortment of used furnishings from their respective homes. Among other items, my father had brought the two Tucson daybeds with their matching floral polyester bed linens, a mishmash of dishes and kitchen essentials, my high school desk, chair, and study lamp, as well as Nanie's Singer treadle sewing machine and a used Electrolux vacuum.

Dick's parents brought an old kitchen table, three chairs, a rug, toaster and assorted kitchen necessities. After the four anticipatory grandparents had arranged everything in the studio as a lovely surprise, they were ready to see us married at five o'clock the following evening before I had even arranged for a night off from work at the library!

Dick had it worse than me. Not only did he have to resign from his fraternity and the ROTC, he had to tell his coach, athletic teammates, and his fraternity brothers that he was getting married due to extremely bad luck in spontaneous semen dispersal.

All I had to do was suffer through the accusing looks of Mrs. Lindstrom, my elderly Lutheran landlady. Saturday morning, when I opened my door and stepped into the hall to use the communal bathroom, I saw a street length ice blue crepe dress with a high princess waist and small bolero jacket hanging on the bannister. A straw wreath with a veil was visible through a plastic sack pinned to the hanger. A pair of matching pumps hung from another hanger in their own sack.

Mrs. Lindstrom came up the stairs looking at me quizzically, and said, "Your mother left these for you early this morning. She said you are getting married today. Is that true?"

"I don't know!" I replied before starting to weep.

Late that afternoon, I was threatened and physically extracted from my rented room by my parents. Before the ceremony, I sobbed in the pastor's office, pleading with my father to not make me marry Dick.

"Please, Daddy, don't make me do this. Please!" I begged.

My father's response was slightly sympathetic. "Pull yourself together now, Joan. Just get it over with. As far as I am concerned you only have to give it a six-month try. The baby will have been born by then, and, if you still feel it is a mistake, you can get a divorce!"

My father was well aware that I did not love Dick enough for marriage. After Dick had written me a "Dear Joan" letter in Europe, I wrote my father on March 14, 1962 about the end of their relationship.

"I am more than a little relieved. I have been holding up my end of our relationship with the kind of effort that is hard to put forth. I have been very honest with him in all ways and this has been rough on him. I do not, nor have I ever, loved him enough for marriage. The only thing I would like to do now is to keep him for a close friend and this will be the hardest thing."

Eventually, I took this prenuptial promise of divorce, written in my father's hand on a blank telephone message torn from a pink pad on the pastor's desk, which was signed by both of us, and placed it beneath my right foot inside my mother-supplied wedding shoe. I entered the Chapel of St. John's from the side door with my head down, without the support of my father's arm, and walked toward Pastor James Comfort Smith and the groom, both of whom had been waiting for over an hour.

It was the first time Dick and I had seen each other since our Thanksgiving coital experience in Tilden Park. I was only vaguely aware of the witnesses. Besides our parents, those present included Dick's sisters, Nonny, and Lary Carpenter, Dick's best friend from EBHS. As if I was swallowing a petrified fried egg, I tried not to look up or feel any kind of sensation until the ceremony was over.

Our barely audible vows spoken, we crossed the Bay Bridge in three cars for an impromptu wedding dinner at Alioto's Restaurant on Fisherman's Wharf. Our party was seated at a rectangular table near the swinging kitchen doors, where a cacophony of Italian men cooking, shouting, and serving saved us from even pseudo-polite conversation. My uncontrollable weeping resumed while my father tried to lighten the mood with humor and conventional social behavior, as if nothing was wrong, by presenting me with some kind of cleaning product in a cardboard box as a bride joke.

The evening of March 23, 1963 still holds first place as the most shameful social experience of my life.

After dinner, my parents drove Dick and me back across the Bay Bridge to a Berkeley studio apartment neither of us had ever seen.

As Dick and I got out of the car, my mother said, "Goodnight, my darlings, sleep tight!" her tone and syntax sounding as if someone had dropped a tincture of Aunt Mary into her Alioto's crab cocktail.

My father escorted us up the stairs and, using the single key, opened the door to the studio, and as he handed the key to Dick, said, "Goodnight and good luck!"

Dick and I had not said anything except our so-called vows to each other all evening. At that moment, we both felt as if our lives were over. I thought of arranged marriages in other cultures where bride and groom are strangers. Ours was a mutant shotgun-arranged marriage. The groom would have been considered a prize catch since his future looked bright—the bride, not so much.

Having inadvertently committed a conceptual sin I had vowed never to let happen, it was now up to me to rectify the situation. Was karma at work for my having committed my mother against her will in Tucson? Had she had taken her retribution, forcing a kind of parallel commitment on me? At that moment I would have much preferred the locked psychiatric ward of the Tucson Medical Center to the Sinners' Barracks on Haste Street.

Tucson Torture Racks in the Sinners' Barracks, Haste Street, Berkeley, March 1963

Inside the Barracks, there was no black lace gown and peignoir set, but instead one in my mother's favorite ice blue color. She had laid both items out on one of the Tucson daybeds with a note.

"I thought you might like to have these for tonight. You'll have to move your other clothes and books from your room at Mrs. Lindstrom's tomorrow. XOXO Mom."

Undressing in the bathroom, I looked at my full naked body in its pregnant form for the first time. The presence of another life was unmistakable. Just as I started to turn away, the fetus inside me stirred and my abdominal wall moved in response. Something deep within my badly battered psyche went "Ping! Ping!" and I felt suddenly filled with instinctual protection for the child inside me.

I blew my nose and washed my face with a bar of my father's favorite Palmolive soap before slowly opening the bathroom door to face the groom. Dick had stripped down to his boxers and t-shirt. His belongings had yet to be retrieved from the fraternity house.

As soon as our eyes met, my tears resumed. I told him how sorry I was and promised I would get him out of the mess as soon as I could.

As if he hadn't heard me, Dick came towards me saying, "I would really like to try making love again. I know that it wasn't good the first time!"

I could not keep myself from laughing, responding, "Dick, as batting averages go, you are batting one thousand!"

My joke broke the terrible tension and we began making fun of our parents. Dick and I both felt as if we had been sentenced to some bizarre immorality prison. The prison barracks had been equipped with two narrow beds; a decorative scheme deliberately designed to constantly remind us of the nature of our singular sexual sin.

After an hour of painful processing, mixed with mutual bouts of hysterical laughter, Dick gently pushed me down on the daybed and climbed on top of me. As he did so, I lost control of my bladder. Was the incontinence from old trauma, new trauma, or the pressure of the new life growing inside me? Whatever it was, our union was not consummated that night, or any other. Damp shame once again defined me.

Those few days in March 1963 were extraordinary. My mother never again seized power in the same way. She was like a fierce mother bear protecting her cub. Her instinctual behavior, combined with her religious and cultural conditioning, and multiplied by the stringent social and reproductive rules of that era, infused my mother with the strength to swat away all obstacles and restrain her most dangerous schizophrenic impulses. Whatever it was, my mother executed a complex and creative plan—one that saved the lives of her only progeny.

BOOK II

MOTHERHOOD

Probably there is nothing in human nature more resonant with charges than the flow of energy between two biologically alike bodies, one of which has lain in amniotic bliss inside the other, one of which has labored to give birth to the other. The materials are here for the deepest mutuality and the most painful estrangement.

Adrienne Rich, Of Woman Born[*]

---

[*] Rich, Adrienne. *Of Woman Born – Motherhood as Experience and Institution.* New York: W.W. Norton, 1986, 220.

-36-

The Love of My Life

I fell in love with the life inside me the same night I started serving my sentence for immorality. The sight and sensation of movement triggered something intrinsic. Commander Margaret was relieved of her post—her diseased morality replaced by worshipful reverence.

As is often the case with prisoners sentenced to serve time in the same cell, Dick and I developed an awkward familial relationship. After my sixth month of pregnancy, and our second month of cohabitation, we found a path through shame and regret to an uneasy normalcy. We reached a business-like agreement that our highest responsibility was to bring the life we had created safely into the world. We were just two UC Berkeley students, accidentally assigned a joint biology project—albeit one of considerable importance, far beyond the National Science Fair.

In addition to our personal course in life science, we shared an intense dislike of the "Sinners' Barracks." Each of us structured our daily schedules to spend as little time in the Haste Street studio as possible. I continued to work as night supervisor at the library, adding additional weekend hours to pay for a new set of needs. Dick replaced his ROTC scholarship income with a few hours of paid work in the Athletic Department. Because we both were carrying a full course load, it was rare for us to be in our tiny cell at the same time. By the time Dick's games and practices ended, I was on my way to work at the library, relieved whenever I returned to find Dick already asleep.

At the end of June, we escaped the Sinners' Barracks and moved into a two bedroom second-floor flat located at the corner of Parker Street and Benvenue Avenue. Anxious to end the agony of the Tucson Torture Racks, we each bought a used bed from departing tenants. While we had no car or television, we each had a private bedroom. My room included a new crib. On my way to see the OBGYN team at the UC Medical Center in San Francisco, I purchased two sets of real linen sheets for ten dollars at an estate sale. They were so fine that even when freshly washed and line-dried, they were as soft as silk.

I was pleasantly surprised to discover that my advanced calculus professor lived in the house next door to our new flat. Because Dick struggled to pronounce her ex-husband's Russian surname, he began addressing her as "Professor Calculus"—nomenclature she seemed to prefer. Until then I had no idea Professor Calculus was a divorcee with five children, who had no time for housekeeping. Because her negatives matched my positives, we found it easy to help each other.

Professor Calculus's first positive was that she lent me a black silk faille maternity dress to wear to Nonny's wedding, which took place on June 15, 1963, the day before my twenty-first birthday. Nonny kept her wedding a secret from me until the invitations had been mailed, afraid to tell me she didn't want me to be her Matron of Dishonor. Instead my role was limited to standing discretely behind a tall podium holding the guest book. Nonny's wedding ceremony was a carefully orchestrated extravaganza, right down to the flowers by San Francisco's top florist, Podesto Balauchi, and the bride's hand-rolled veil. While the bride was not virginal, neither was she pregnant. She was stunningly beautiful. I cried for both of us.

Nonny and Jim, Bride and Groom, June 15, 1963

My August 17 due date was not difficult to establish, since there was only one possible hour of conception. On August 16, I had an appointment with the OB-GYN resident who examined me through the last four months of my pregnancy. As I lay on the exam table, my feet in the humiliating stirrups, the male resident in charge of my case narrated his findings before making an announcement to the surrounding interns.

"She isn't even effaced; it will be several weeks before she goes into labor," he said authoritatively.

Waddling out of the UC Medical Center, I realized there might not be another opening in my schedule for a very long time. I decided to use the remaining hours before my library shift to relax. Following a craving for crab, I walked down to Fisherman's Wharf and wandered the stalls of vendors with a paper cup of crabmeat in hand. After a second crab treat, I awkwardly climbed aboard a trolley to Union Square, where I took a seat on a sunny bench near several destitute schizophrenic ladies. Those disheveled mumbling women stood in sharp contrast to my mother, who had been labeled by her Tucson psychiatrist as a "high functioning" paranoid schizophrenic.

Warmed by the sun, I bought an ice cream and window shopped as if I was back in Paris, free to wander the streets. Boarding the four o'clock bus to return to Berkeley, I felt a strange wave of nausea and cramping. Maybe the crabmeat was spoiled? By the time the bus had crossed the Bay Bridge, my water had broken and I was in active labor. The driver and a passenger kindly helped me into the backseat of an Oakland policeman's patrol car who drove me to the nearest emergency care—Providence Hospital.

Confident and unafraid, having assisted with natural childbirth in Africa, I believed giving birth was something I could handle with ease. If anything, I was not so much in a state of active labor, as in a state of active arrogance. As I lay on a gurney in the labor room, my Frank-V.-Lieberman-trained-eyes noticed hundreds of hand prints on the wall next to the gurney. Distracted by needed maintenance, I began pondering their source. Suddenly a strong contraction sent my own hand slapping against the wall, where it left an imprint as I rolled over in search of an escape from pain. Then I felt frightened.

I gave birth to a baby girl twenty-three hours later. She was right on time—behavior that still persists. I looked into her eyes and was gobsmacked. She was the most beautiful human being I had ever seen. Love for her poured out of my heart and filled my breasts. Blackie never even crossed my mind.

Commander Margaret made one last surprise appearance on the day I gave birth, having flown to San Francisco without asking me or telling my father. When no one answered her knock at the Parker Street duplex, my mother went to the library to learn my whereabouts. I came out of a post-delivery nap, opened my eyes, and, bingo, there was my mother!

Silently I screamed, "Help! I need the boundary patrol!" until I remembered my daughter's face and my Pavlovian panic subsided.

Dick didn't meet his daughter until the following day. Returning late from a pre-season Cal Bears football scrimmage, he had gone to bed assuming I was at work. In the morning, I woke him with the news.

"Dick, we won first place with our biology project," I exclaimed. "The baby is a girl and she is beautiful!"

Olivia at two months with her biological father, Dick Alton, Berkeley, October 1963

My daughter taught me about love. Up until her birth, I had lived without having fallen deeply in love with another human being and so I had missed the most fundamental gift of life. When we feel true love, it is magic meeting truth.

From the moment we are born, we are instinctively focused on receiving the love we need to survive. By the time we wake up from childhood, our psyches have been conditioned to focus on how much love we are receiving, carefully noting rewarded behaviors. Each of us has a love deficit. If the deficit grows disproportionately large, we may forget how to love others as a defense.

When the unfathomable depth of my love for my newborn daughter made a bulls-eye strike, it knocked me into another reality. Looking into her eyes, I was filled with a delicious falling-in-love stew—a blissful mixture of elation, tenderness, awe, and devotion.

When I called my father collect to tell him he had a granddaughter named Olivia, his response confused me.

"I'm glad it's a girl," he said. "We don't know anything about raising boys!" as if we were going to do it together.

Dick's parents generously gave us a washing machine as a baby present. Three months later, on November 22, the Thanksgiving anniversary date of Olivia's conception, I was on the back balcony of the Parker Street duplex hanging diapers on a pulley clothesline, humming along with a new Beatles' tune on the radio. Suddenly the music stopped mid-song—President John F. Kennedy had been shot in Dallas.

Dick and I carried Olivia next door many times to watch the aftermath of President Kennedy's assassination on Professor Calculus's television set. On Thanksgiving Day 1963, instead of a pre-coital fried chicken picnic in Tilden Park, Dick and I nibbled on celery sticks with peanut butter and Stouffer's macaroni and cheese with Professor Calculus and her children. None of us could take our eyes off the television long enough to shop or cook.

That fall Dick stayed with Olivia while I worked the library night shift. He learned how to change diapers, prepare a bottle of frozen breast milk, play peek-a-boo games with her, and toss her about in the air as men often do with babies. But it was rare for Dick to take Olivia out for a walk, nor did he invite any of his friends or teammates to our flat to show off his beautiful daughter.

On nights when the Cal Bears required Dick's presence, Rebecca, the eldest daughter of Professor Calculus, stepped to the plate as his pinch hitter. A serious student in her final year at Berkeley High School, Rebecca was highly experienced, having helped with the care of her four younger siblings. Also, her mother was next door if there were any problems. In exchange for her assistance,

I cleaned their house, did the family laundry, and paid Rebecca a small hourly wage. She claimed she liked my services more than my money.

My mother stayed on Thorazine through my pregnancy, but after Olivia's birth she began to secretly stash her pills away. In December, Dick and I took Olivia on the Greyhound Bus to Bakersfield for Christmas with his family. The day after Christmas our trio boarded another Greyhound to Tucson. During the long hours on those buses, I saw more fully how the birth of a child changed everything—particularly how both Dick and I wanted our parents to know and love Olivia.

Upon arriving in Tucson we discovered my mother had been gone for two days. According to my father, she had been disappearing with increasing frequency. Most of the time she was across the border in Nogales, hiding from the FBI. My father's reaction was within his normal range, limited to leaving her psychiatrist a message.

Still in a state of denial, he said, "Margaret is out of town so she won't be coming for her regularly scheduled appointment."

Dick and I slept on new twin Tucson daybeds and agreed they provided small increments of comfort over the originals. When I got up at dawn to nurse Olivia, my mother's purse was on the kitchen table. She had made Olivia a Raggedy Ann doll. At the sight of it, I began weeping—feeling old shame for having defecated under the lilac bush, shame that still could not be scrubbed away. The Raggedy Ann doll my mother had been making for my fifth birthday was instead given to Marlene for her seventh a month later.

Before we climbed aboard the Greyhound Bus for the long ride back to Berkeley, I called my mother's psychiatrist and told him that his patient was clearly off her medication. But Olivia's birth had increased the distance between my mother and me. Now Olivia was my highest priority. While scary memories still lurked, I had new responsibilities—ones infinitely more compelling and more joyful. Dr. Weinberg's invaluable prescription to create a new life for myself had been filled and the treatment was working.

Olivia celebrating her fifth month of life, with her mother, Berkeley, January 17, 1964

-37-

Excommunication and the Ayn Rand Man

My parents had been without a couch since their move to Tucson. My mother was in such poor mental health at the time of the Tucson transfer, she once again believed the suffocating snakes were still inside the well-worn cushions of our amputated, re-upholstered couch. When the moving men from Mayflower arrived on Hollins Street, my father told them not to load it. My parents didn't buy another couch during the remainder of their marriage. Separate chairs provided more tolerable distance.

The Tucson Torture Racks were the first so-called couches in my own household. Dick and I reluctantly kept them in the living-dining room of the Parker Street duplex. I was sitting on one while nursing Olivia when I asked Dick for a formal pardon from our shotgun marriage. Not only had I already consulted a lawyer, I had worked very hard to pay the lawyer's low-trust advance fee of three hundred dollars.

While our mutant shotgun marriage had been given an unplanned four-month extension, I felt that Dick had suffered long enough. I was shocked at his visible hurt when I asked him to move out. All his previous signals indicated he was anxious to return to his pre-marital life. Dick carried away his own bed and linens, my high school desk and chair, along with a few useful hot plate items. I was left with the rest of our second-hand household effects, including the Tucson Torture Racks.

Ironically, to legally end our marriage, Dick had to be the plaintiff in the proceedings and I had to agree to be labeled an adulteress—then the only grounds for divorce available under California law. After Dick moved out, our collaborative parental relationship slowly disintegrated. Having served our joint sentence in immorality prison, not only did I assume the fiscal and social burdens required to unlock the legal handcuffs holding us together, but only I remained on perpetual probation.

I am still deeply ashamed to admit that I did not worry about how Olivia would experience Dick's absence. My concerns were structural. Dick promised he would continue to provide care for Olivia while I worked at night, but he always seemed to keep a self-protective distance from his daughter. He was not a proactive parent and I was blind to the potential damage to Olivia. Given the still strict social norms of the era, Dick had done what Commander Margaret demanded and society expected, but it was a barely believable performance. Dick never seemed fully present to me. Whenever I asked him what he was thinking, he responded with a sports report.

For the first year of Olivia's life, the only in-person contact we had with my parents were right after her birth and our post-Christmas visit to Tucson. My father usually wrote me twice a month, his preferred method of communication, providing his kind of *optimal distance*. His letters were benign narratives of weather, crop disasters, grocery and gas prices, descriptions of his latest research projects, and changes in lab personnel, particularly stenographers. My mother's letters covered a different set of topics—ranging from the FBI's plot to destroy her documents to the coming international devaluation of paper money, mixed with obscure Biblical references, most written in shorthand.

Olivia in Berkeley's Tilden Park, Thanksgiving 1964

For Olivia's second Thanksgiving, I took her on a picnic in Tilden Park with other friends. She wore my sweater which I had deliberately shrunk and cut the sleeves on to keep her warm.

In early January Dick asked to take Olivia to Bakersfield to visit his parents for a few days. I felt unable to refuse since Olivia had weaned herself, and, at seventeen months, was enjoying a wide variety of foods. But when Olivia developed a temperature while away, I boarded a Greyhound Bus to Bakersfield to care for her. By the time I arrived, her temperature was normal and my emergency medical mothering was no longer needed. Dick decided to extend his Bakersfield visit, but I needed to return to work. Lary Carpenter, who had witnessed our marriage, kindly offered to drive me and Olivia back to Berkeley.

The five-hour drive to Berkeley was the beginning of a deeper friendship between Lary and me. Perhaps my latent car hormones directed my affections—Lary had a sea green MG convertible. During the time Dick and I had lived together, Lary had stayed with us several times when he visited Berkeley to attempt to enroll in the College of Environmental Design. At EBHS, Lary had been a champion wrestler, not a champion scholar. But after two years at Bakersfield Junior College, Lary had finally met the eligibility requirements for UC Berkeley.

Lary had the muscular body of a wrestler and stood five feet ten inches tall with straight brown hair. Because he suffered from terrible astigmatism, his hazel eyes were always hidden behind ebony-framed glasses with lenses as thick as the bottom of a Coca-Cola bottle. Unlike Dick, Lary's interest in sports ended after high school—perhaps because he became rebellious against all things normative. He had begun smoking a pipe with obvious affectation and ·spoke with a soft, solemn voice—half pretentious and half awkwardly shy.

Lary was in the midst of developing a life philosophy. He had read most of Ayn Rand's work;[30] her novel, *The Fountainhead*, was the source of his motivation to become an architect. Lary had also adopted Rand's belief that the proper moral purpose of one's life was the pursuit of one's own happiness, justified by what Rand called "rational self-interest." Lary wanted a society that fostered and supported individual rights and creativity. Since my own idealism was deeply attached to the common good, we were at philosophical odds.

As the sea green MG moved the three of us north, Lary and I argued about our diametrically opposed worldviews. Lary had also decided he was an atheist, a decision terribly upsetting to his mother. I was agnostic, so we spoke about the differences in our religious upbringings. My discussions about God with Lary reminded me of an important task on my crowded must-do list. It had been three years since I had vowed to have my name removed from the membership list of the Mormon Church—a vow made in March 1962, at *Foyer de Bonté* after the two Mormon missionaries arrived to dissuade me from going to Africa. After returning to Berkeley, I wrote a letter to the bishop of the University Ward, drawing on the kind of arrogance and unauthorized practice of medicine that characterized my young adulthood.

> *March 5, 1965*
> *Bishop Marc Ricks*
> *The Church of Jesus Christ of Latter Day Saints*
> *University Ward Bishopric, Oakland-Berkeley Stake*
> *Berkeley, California*
>
> *Dear Bishop Ricks:*
>
> *I am writing to request excommunication from the Mormon Church on the grounds of apostasy. I was baptized in the Logan Temple in July 1950. I have not been active since 1956. My parents did not raise me in the Mormon faith. My baptism and participation were driven by my own childish needs for safety and inclusion, not by faith.*
>
> *After years of reflection, I have come to the conclusion that Joseph Smith was neither a prophet, nor an instrument of God. Rather, while he had a high intelligence quotient and considerable charisma, Joseph Smith was most likely someone who suffered from manic depression—a mental illness often characterized by grandiosity, hallucinations, hyper-sexuality, impulsiveness, creativity, and risk taking.*
>
> *Please let me know if there are additional steps that I need to take in order to assure my name is removed from the membership list of the Church of Jesus Christ of Latter Day Saints*
> > *Joan Carol Lieberman*

Excommunication required a "Bishop's trial," a proceeding subsequently made famous by Sonia Johnson during the nation-wide political battle to pass the Equal Rights Amendment, which the Mormon Church opposed.[31] During my excommunication trial, I was alone, without counsel or companionship. Bishop Ricks and two other bishops ruled that because I had been blessed to know and accept the gospel of the Church of Jesus Christ of Latter Day Saints, but was now rejecting LDS principles, I was condemned to burn in hell for eternity. Emotionally battered by the Bishops, I was nonetheless completely cured of organized religion. After the trial, I vowed to continue to tithe—but not to the Mormon Church. Instead, I would give ten percent of my earnings freely to those in need because giving had taken root in my soul independent of any doctrinal praecipe.

Notice of a thirty percent increase in rent on the Parker Street duplex made it necessary for me to find a new place to live. Lary was exploring the Berkeley rental market, and since I had no time or car with which to do so, I asked him to keep an eye out for a place for Olivia and me.

A few days later, calling me by my initials, as was Lary's habit, and demonstrating Ayn Rand's rational self-interest, he said, "J.C., I think we should look for a place to live together. It would be good for both of us in a lot of different ways."

We moved into a ground floor duplex at 2732 Derby Street, equal distance between my old single room at Mrs. Lindstrom's and St. John's Presbyterian Church, the site of my shotgun wedding ceremony. The monthly rent on the Derby Street duplex was $75 less than Parker Street, but more importantly it also included a fenced yard for Olivia, a small back room for Nanie's Singer sewing machine, and plumbing hook-ups for the Alton washing machine.

Lary and I finished moving the last load of my furniture in a rented pickup truck and dropped the Tucson Torture Racks at a free furniture site in West Oakland. As we climbed back into the truck, Lary announced he wanted to drive to Chinatown in San Francisco to hunt for wooden packing crates. It was nine o'clock at night and Olivia was asleep on my lap, but Lary had an idea about new architectural couches made of crates. As we drove toward San Francisco, I remember thinking that it might be best if I ended our joint tenancy agreement before it was too late.

I soon discovered that living with Lary was like being drawn into an ill-conceived dramatization of Ayn Rand's *The Fountainhead*. Wanting to embody the individualism of Howard Roark, Lary's creative design ideas dominated our shared duplex. He painted the dining room Chinese red, the bedroom dark gold, and Olivia's room green, as in the color of healthy grass. In the open areas of the living and dining room, Lary created unique special arrangements of the Chinese crates. He "upholstered" the crates using foam rubber packing material and burlap, dyed with Ritz colors in the washing machine, tacking the burlap over the packing material with rough, hand-forged nails. While I was out working to pay the rent and buy food, Lary dyed single panel see-though burlap blinds for all the windows, leaving Ritz residue in the washer so that most of our clothes became color-coordinated with the decor.

Lary Carpenter on his Chinese Packing Crate "Couches," Berkeley, 1965

When I complained to Professor Calculus that Lary's style of decorating was something between "Berkeley drug den" and "too poor and stupid to know better," she responded that Rebecca thought Lary's interior design work was newly chic. Granted, his creative crate couch arrangements were developmentally right for Olivia—she loved to play hide and seek among them. Still, I recall only a few days I didn't have to put her through the discomfort of splinter removal.

Decorating style aside, there was an invaluable over-riding benefit of living with Lary. He had either inherited or absorbed the excellent teaching and coaching skills of his father, Les Carpenter, who had been Dick's high school baseball coach and my guidance counselor at EBHS. Lary became devoted to Olivia—he was kind, tolerant, and protective of her. In this respect, Lary more than made up for what had gone missing when Dick disappeared from her orbit.

Consistent with Rand's philosophy of self-responsibility, Lary taught Olivia skills that most parents of children her age would not have risked, such as how to safely cut vegetables and bread with a sharp knife and to fold and put away her own clothes. While building a loft bed and a child-sized table for Olivia, Lary taught her how to use a hammer, saw, and other tools.

Lary's shy authoritativeness made his guidance more powerful and Olivia responded with affectionate obedience. After he told her she was too old for a nighttime diaper, Olivia immediately refused to wear one. She went the rest of her childhood without waking up in a wet bed—and from her mother's late incontinence to her own determined continence in one generation.

Olivia cutting lettuce for Lary, Derby Street duplex, Berkeley, September 1966

Watching Lary's care and guidance of Olivia increased my affection for him. Whether it was Ayn Rand or the larger cultural context, Lary was deeply engaged in developing a philosophy for living and was willing to discuss the contradictions of various life choices openly with me. A mixture of his devotion to Olivia, his intellectual vulnerability and vitality, combined with his powerful sexual presence, slowly expanded our relationship into a romantic one.

As for Dick, he understandably experienced the relationship between Lary and me as a double betrayal. Dick had other reasons to be angry, but my relationship with Lary motivated him to put even more distance between himself and fatherhood. He slowly disappeared from Olivia's life. Dick paid the court-ordered monthly child support only twice; then, just like our first date, I said I had money enough to pay for his share. Soon Dick was back in full swing with his old fraternity brothers and sports.

While I no longer had to masquerade through Mormonism, I still felt split—divided between an increasing sense of hope that, at age twenty-three, I might somehow escape the schizophrenic executioner and the steady drum beat in medical journals condemning me to my mother's fate. Only Dr. Fox's powerful argument against those theorists gave me hope.

-38-

When the Body Remembers

Berkeley was in such a state of upheaval that whenever I pushed Olivia's stroller towards campus, we entered another reality. It is sometimes hard to remember that Mario Savio was wearing a suit and tie in September 1964 as he stood on the second floor balcony of Sproul Hall urging passing students to join the Free Speech Movement (FSM). Only a few months passed before the majority of those seen on the streets of Berkeley were anti-war demonstrators, street people, or hippy "flower children." Shop windows on Telegraph Avenue were boarded up or had newly installed metal bars as Berkeley became enveloped in the chaos of cultural change.

It was an economic necessity for me to return to work three days after Olivia's birth. At the end of the first week of working-nursing-student-motherhood, I was forced to face the impossibility of being a mother, earning a living, and keeping up with UC Berkeley's rigorous pre-med program. I formally abandoned my dream of becoming a doctor and withdrew from the pre-med track.

For two years I had been supervising forty part-time library employees, each of whom seemed to suffer myriad emergencies. After the head librarian praised the team system I created for scheduling flexible work shifts, I decided to study organizational management and transferred to the Institute of Industrial Relations.

My management role provided a front-row seat to the daily drama created by Berkeley's Free Speech Movement (FSM), which had a visible impact on the organizational behavior of students, faculty, and staff. FSM leaders had called for a general strike of both faculty and students in early December 1964. The unionized staff of the library met to take a strike vote. The librarians faced the same kind of choices as the teaching faculty, the majority of whom cancelled their classes.

As a student who was also part of management, I watched these debates closely and noted the unintended consequences every time UC Berkeley administrators tightened campus rules. For a class on labor-management relations, I wrote a paper describing what I had observed in the library and elsewhere. The professor encouraged me to submit my paper to an academic journal for publication, but first he asked my permission to send it to President Clark Kerr.

A few weeks later, the President's secretary sent me a note—President Kerr wanted to meet with me to discuss my paper. Lary drove me to the Kerr home in the hills of El Cerrito, waiting outside with Olivia until the meeting was over. While his wife, Kay, served us coffee and cookies, President Kerr peppered me with questions, encouraging me to expand and publish the article. While flattered and intellectually stimulated by his interest, I had no idea then that President Clark Kerr would play a important role in my professional future.

Lary's period of study in the School of Environmental Design also came at a time when the curriculum was in the midst of reformation. Creative thinkers and designers like Sim Van der Ryn and Charles Moore were re-defining the role of architects. Then one wave of architectural creativity accidentally led me into a memory space that almost ended my life.

Some of Lary's fellow Environmental Design students discovered a source of free building materials: the old timbers and driftwood that perpetually washed out of the San Francisco Bay onto the Emeryville Mud Flats. In addition to creating sculptures on the Mud Flats, students were using the flotsam to build lofts and furniture. Lary decided he was going to make a "real couch" having experienced the discomforting limits of his artistically arranged Chinese crates.

On a Saturday morning, we drove to the Emeryville Mud Flats in a borrowed truck. According to my diary from that era it was February 20, 1966. Dick had taken Olivia to spend the weekend with his parents in San Francisco. Olivia was very excited because, in addition to going to the Zoo and Golden Gate Park, she was going to spend Saturday and Sunday night with Grandma Florence and Grandpa Harry in their hotel.

At the Emeryville Mud Flats, the strong wind coming off San Francisco Bay filled the air with stinging particles of sand and the noxious smell of rotting garbage mixed with decaying plant material. As I stood in the mud, surrounded by piles of broken branches and driftwood that students and artists had turned into eclectic sculptures, I lost myself. The sensation of nausea came first, followed by blinding dizziness. Feeling faint, I sat, then lay down. When I was able to get up on my knees, I tried to crawl into the Bay. When Lary noticed my absence, he quickly pulled me from the water and helped me into the truck.

I wanted to die rather than continue to feel such overwhelming distress. In episodic moments of lucidity, I felt certain I had finally been stricken with schizophrenia. Later that night, while Lary was in the shower, I left our Derby Street duplex and somehow began crawling up Ashby Avenue. I woke up in Alta Bates Hospital, where a clinical psychologist named Kenwood Francis Bartelme was in charge of my care. Dr. Bartelme, tall and lean, with dark hair, was a former Merchant Marine. He understood the vicissitudes of war and the after-effects of trauma.

When he came into my hospital room, all he said was, "I see nothing in your tests or your behavior to indicate that you are suffering from schizophrenia yet."

Dr. Bartelme had no reason to say anything else to me because the psychiatric community was still in the middle of the era of the "Schizophrenogenic Mother." However, my case was interesting to Dr. Bartelme because my behavior was similar to that of several young Vietnam veterans under his care. Since I had never taken any psychedelic drugs, he thought I had somehow triggered a memory of previous trauma. I don't recall we ever attempted to determine its exact nature.* I was as familiar with the current literature on schizophrenia as Dr. Bartelme, having developed the habit of visiting the Periodical Room during my library breaks to read the latest research articles.

Olivia was my source of hope—her joy in life was contagious. Still, my breakdown on the Emeryville Mud Flats was a real setback. The immunity created by my love for Olivia had been weakened. In a state of heightened hyper-vigilance, I once again feared that I was on the precipice of schizophrenia. Dick's mother, Florence, remained at the top of my substitution list. Even though I hadn't yet asked her, I had faith that Florence would gather my golden baby chick under her wing and keep her safe and happy.

Dr. Bartelme wrote my discharge orders and gave me an appointment to see him two days later. Professor Calculus signed my release and drove me home. As I opened the front door, Olivia ran into my open arms. Then taking my hand, she led me into the front room.

---

*I subsequently came to wonder if the scent of decay and the stick figures on the Emeryville Mudflats triggered my unconscious memories of "The Day the Bear Went to Topaz"—it appears that I experienced a kind of time collapse as the events occurred on the same day, twenty years apart.

"Look!" she spoke with pride. "I helped Lary pull the big needle to sew the cushion on our new couch!"

The back support of the new couch was a piece of timber, three times the size of a railroad tie, and the bottom was more of the same. Even though the new structure was an improvement over the Tucson Torture Racks and the carefully configured Chinese crates, I was sane enough to know we were still a considerable distance from real comfort.

Olivia on porch of Derby Street Duplex, March 1965.

-39-

Vietnam War Bride

The war in Vietnam was tearing our country apart. Lary's draft lottery number meant it was likely he would soon be on his way to Vietnam. After we learned of the combat deaths of two friends in the same week, Lary couldn't sleep. His hands tremored lifting his coffee cup to his mouth.

Just as I had in Wickel's Men's Store, when I saw the hands of a man begin to shake, I impulsively volunteered the only form of protection I had to offer. I volunteered to save Lary from the draft by marrying him, providing him with an instant exemption—a wife and a dependent. My perfunctory offer was without social sacrifice since my value as a bride was gone. I was definitely not a virgin and my divorce from Dick had labeled me an adulteress. Also, in 1959, after reading "The Perverse Mother" in the library of Bakersfield Junior College, I had forsworn marriage. Lary accepted my proposal.

My second marriage ceremony—voluntary, but with some symbolic shotgun imagery—took place at dawn on March 27, 1966 in the backyard of the Derby Street duplex. I forgot March 27 was the date of my parents' marriage. UC Berkeley Professor Fred Sheridan Stripp performed the ceremony. A well-known UC Professor of Rhetoric, he also happened to be a Congregational Minister, and was running for mayor of Berkeley on an open-housing, anti-war platform. Professor Stripp had given me an A+ on my final oral presentation— likely a late reward for my having sat through all those Mormon Testimony meetings in Logan.

I staggered out of bed, pulled on a crocheted sweater and wool knit skirt, (the same articles of clothing I had worn to work a few hours earlier), and washed my face. Then I awakened Olivia, who held her infant rubber doll "Bookie" through the five-minute ceremony. Like my first marriage, Nonny also witnessed my second—this time joined by Jim. Nonny surprised us by baking a beautiful wedding cake, decorated with fresh flowers "borrowed" from the UC Berkeley grounds. Since the ceremony was over by six o'clock in the morning, the six of us ate Nonny's delicious cake for breakfast with scrambled eggs. The snapshots show a puffy-eyed bride and sleepy child. The early ceremony was Lary's idea; dawn has never been my finest hour.

Vietnam War Groom and aride, six o'clock in the morning, Berkeley, March 27, 1966.
Lary Carpenter, the author, Olivia, and Professor Fred Sheridan Stripp

Lary's parents, as well as my own, were "grievously disappointed" to learn of our marriage. None of our four parents understood that, although there was love and appreciation in our duplex, the relationship between Lary and me was not really marital. Eventually I recognized Lary Carpenter provided me with some of my mother's unpredictability. Further, by keeping a good distance between himself and any fiscal responsibilities, Lary made it possible for me to habitually over-function and pay more-than-my-full share, providing me with a false sense of being in control.

Six months before the marriage, I had begun putting my language skills to use. Professor Sim Van der Ryn, Lary's architecture professor, gave my name to Spiro Kostof, a newly appointed professor of architectural history, who was born in Turkey. Professor Kostof was writing a book about the history of architecture, including Turkey's Cappadocia cave dwellings, for which he needed translations of articles from French and German archaeology journals. It was intellectually engaging work, paid a high hourly rate, plus I could do most of the work at home with Olivia. Ayse Taner, the AFS Turkish exchange student, who had kept a culturally inappropriate distance from my father by washing his feet and anointing them with rose oil, dominated my dreams for weeks.

While I was mothering, studying, supervising, and translating, Lary became sexually involved with another student, Anne, a beauty from Beverly Hills, who had her own MG (white), and a private apartment on the north side of campus. Lary met Anne in Dwinelle Plaza during an anti-war protest rally.

The sudden absence of Lary from our shared duplex ate away at my self-esteem. I struggled to persuade myself that providing Lary with an exemption from the War in Vietnam had been a reasonable response to a corrupt war, not proof that I was so easily disposable. Yet, on the days when I was premenstrual, I felt not only that I had thrown my body on a grenade to save Lary, but that he had reciprocated by putting his foot on my back to ensure the trigger went off.

-40-

Mistletoe and Modern Day Polygamy

If I needed Dr. Bartelme, Berkeley needed him even more. The community was split, as if Berkeley was the focal point of the dialectics then at the forefront of America's psyche—between war and peace, freedom and restraint, love and abstinence. Boundaries were dissolving. The distances between what was lawful and what was unlawful, between self and sacrifice, and between morality and immorality, were in flux. Then a different disturbance appeared on our psychedelic panorama. This one came with special meaning for me—after all, what are friends for?

Two years after their marriage, Nonny and Jim had withdrawn from San Francisco State College and moved into an apartment on the north side of Berkeley. Jim transferred to UC Berkeley and Nonny enrolled at the California College of Arts and Crafts in Oakland. Because of their new proximity, we began to socialize as couples. Jim and Lary each had their artistic interests, Lary's in architecture and music and Jim's in poetry and music. Not only was there considerable conversational material, we all benefitted from Nonny's great cooking.

The day after Labor Day, Lary lent me the keys to his MG so I could pick up groceries while Olivia took her nap. I spotted Nonny waiting for a bus on Telegraph Avenue and pulled over.

"Hi, Nonny," I said. "Where are you headed?"

Nonny climbed in before I could offer her a ride.

"I'm on my way to Bill's apartment. Could you drive me there? I'm late and I need to deliver his birthday present!" she said, her voice urgent.

Nonny was carrying a small waxed paper bag and her large straw tote. She began directing me to Bill's apartment on MacArthur Street in Oakland. Because she seemed unusually preoccupied, I did not ask her who Bill was or why she was giving him a birthday present, especially since it was only six doughnuts from King Pin. Also, gift-giving had always been high on Nonny's must-do list.

A week later, I was in our backyard hanging laundry when Jim appeared. With tears in his eyes, he asked to raid my emergency fund: a quart jar I tried to keep full of coins. Handing Jim the jar, his facial expression was worrisome.

"Jim, I can see something is wrong. Is there anything I can do?" I asked.

Silently signaling "No" with his head, Jim turned and walked away, holding the jar child-like, close to his chest.

After dark, Jim returned. He and Lary talked privately on the back steps, while I read and snuggled Olivia to sleep. They had consumed a six-pack of beer before I joined the conversation, but Lary quickly brought me up to date. Nonny had been at Bill's apartment since the previous night. Jim had used my emergency fund to buy sleeping pills and a box of shells for his childhood rifle. Since Jim was currently unarmed, I agreed to show him where Bill lived.

Lary stayed with Olivia, while I chauffeured Jim to Bill's apartment and pointed to the door I had seen Nonny enter. It was a long time before Jim reappeared, pulling Nonny behind him. She appeared to have taken some kind of sleep medication herself. Positioning himself in the passenger seat, Jim pulled Nonny into his lap.

"Joan, please drive us home!" he said, after which no more words were spoken; the tension was terrifying.

When the sun came up again, Jim arrived mid-breakfast. He gobbled the last piece of Bisquick coffee cake before licking the pan clean. Then he announced that he was going to take Nonny to Mexico—not for a divorce, but to keep her away from Bill.

Jim wanted to ask us for a favor, "I was hoping you guys would help me pack up our apartment."

I prefer to do my own packing and had no time to do Nonny's, so I inquired about her whereabouts. Tears from Jim's large brown eyes dampened the front of his blue shirt. I pushed a box of Kleenex toward him. After blowing his nose several times, he told us that Nonny had gone back to Bill's apartment.

"She insisted on being with Bill one more time before she would go away with me!" he said, his normal baritone tone reduced to squeaking.

I was sure Mexico was a terrible idea—it sounded like something my mother would do. Nonetheless, Lary and I complied with Jim's packing request. Postponing our studies, we worked as a silent trio, while Olivia made box houses and drew on packing paper. Each adult was moving without enthusiasm, trying, for different reasons, not to accomplish what none of us wanted to do. By nightfall, Nonny's yellow curtains had been taken down from the windows and the world outside appeared black. Olivia, her nose to the cold glass, reported we were being watched by people in other apartments.

Hunger led us back to our duplex. Instead of dinner conversation, we three adults made a pretense of listening with Olivia to a new album of children's music sent by Florence, while our adult imaginations were riveted on Nonny and Bill. I made a bed for Jim on the sewing-laundry room floor. Then we drifted into dramatically different dreams to the sound of the voice of Johnny Carson on the television of our elderly hearing-impaired neighbor.

None of us went to class. Jim blubbered on the back steps, while Lary offered Rand-restrained comfort. Nonny arrived mid-afternoon, moving like a female panther after a successful kill. At the sight of her, Jim's sobbing increased. It was a warm day, but Nonny wore her suede jacket belted tightly around her tiny waist.

"Stop it!" she said to her husband, before hissing a command, "Lary, please give us a ride home!"

Nonny hoisted her skirt above her knees showing her lush thighs, as she climbed into the MG jump seat, followed by the architect and poet. After their departure, I sat on the back steps watching Olivia pull the petals off a calendula daisy. I wondered how Nonny would feel when she saw her sunny curtains had been stripped away. After a perfect wedding ceremony and three years of marriage, would my best friend regret that nothing would ever be the same again? The pain of her sexual betrayal might eventually lessen, but I believed that the memory of it would change the distance between her and Jim forever.

At midnight, the following night, Jim arrived at our door, trying to keep a stiff upper lip as he announced his wife had just boarded a Greyhound Bus to Bakersfield for a time-out with her mother. Then he added a critical new factoid in their marital drama: Nonny was very proud of having achieved multiple orgasms with Bill, a phenomenon heretofore absent from her marital relations with Jim.

Then Jim's manhood began melting, as he told us, "Even though Nonny might have previously considered going to Mexico with me, she now feels certain she is pregnant with Bill's baby!"

Having rewritten his own part in the script in light of these declarations by Panther Woman, instead of Mexico, Jim was planning to move to Montana to work on a cattle ranch. He asked us for temporary refuge, so we moved their household furnishings into our front hallway. Our duplex soon began to smell of Nonny's dusting powder, spilled during the move—the scent magnifying her absence.

Two weeks passed, as tension increased. Then late one night I answered the telephone and heard Nonny speak without prelude.

"My mother is driving me nuts!" Nonny then asked what I thought Jim would think if she came back to Berkeley even though, she explained, "I've got to wait seven more days before a pregnancy test will confirm I am carrying Bill's child."

As a disqualified marriage counselor, I told Panther Woman to come back. The next day, I stayed on campus until mid-afternoon, trying to give the fractured couple time to work things out, even though Olivia was in serious need of a nap. When I finally pushed Olivia home, I was greeted by faces of failure. Asking no questions, I put Olivia to sleep and started preparing dinner. Nonny napped on our bed, while Jim and Lary took a joint hike in Tilden Park. Dinner was really tense. When Nonny began using her bread to wipe the remaining salad dressing from our wooden salad bowl, Jim exploded.

"You're a fat whore!" he shouted, before noisily stomping out the front door, slamming it hard.

Lary quickly took Olivia out the back door to catch moths, while Nonny and I cleared the dinner dishes. Nonny dried the plates and glasses as if she were stroking Bill's body, while giving a rapturous description of their newborn relationship, telling me that Bill had pictures of her on his wall. Also, on the day she had brought Bill his birthday donuts, he had watched her come across the street with the sunlight shining through her cotton voile dress and he had greeted her from the top of the stairs, telling her she had an extraordinarily beautiful body.

As I cleaned the counters, Nonny explained, "If I am carrying Bill's child," she added, "Bill and I will probably move to Alaska."

Before crawling into bed, I took a long shower in a futile attempt to wash away the residue of betrayal. Lary's relationship with Anne had tainted the already compromised distance between us. Months before Nonny's affair with Bill, I had doubled that distance by agreeing to have dinner at the Hotel Claremont with Dr. Fox. In keeping the openly militarized norms Lary and I had already established, Dr. Fox was not a secret.

Nonny's infatuation with Bill and the possibility of another accidental pregnancy had created a chasm between her and Jim. Philosophically conflicted,

I thought Nonny should have kept her orgasms secret. Yet, while drying dishes, she had confirmed that her love for Jim was ongoing. It seemed we were two California couples engaged in variations of modern day polygamy, a form that proliferated during the 1960s and continued into the 1970s.

As the week of waiting progressed, Nonny was under more surveillance than even my mother could have conjured up. Jim, Lary, and I watched her with hawk eyes, believing she would try to see Bill by making some casual excuse for leaving the duplex. Over the next seven days Jim traced his planned route to Montana on five different maps, leaving them open in the places Nonny most often came to rest: near the cookie jar, on top of the magazine basket by the toilet, on the front porch swing where she painted her toenails, and in Olivia's room where Nonny spent hours drawing hearts and exotic flowers.

Olivia whispered a complaint, "Nonny is using up all my crayons!"

An afternoon telephone call confirmed Nonny's pregnancy test was positive. Jim boarded the bus to San Mateo to tell his parents his marriage was over. Nonny caught the bus to Oakland on College Avenue to tell Bill.

After what felt like a way-too-short intermission in this unscripted real-life drama taking place on my home stage, I heard Nonny before she opened our front door. In a high screech, not much different than Olivia would emit after a knee-scraping fall, Nonny was wailing at full volume.

"A new guy is living in Bill's apartment! He said Bill moved to New York last week!" she spoke through sobs.

It seemed the new tenant had also asked whether Nonny would possibly want the two kittens Bill had left behind. She had Bill's kittens inside her big straw tote.

At the sight of them, Olivia was ecstatic. Before I left for work Olivia had named them "Spotty" and "Stripey." Spotty, the male, had spots of black and white; Stripey, the female, could have been Norrie's baby – their coats were identical.

Over the next ten days, a group consensus developed that Nonny should have an abortion. Jim was failing the course on forgiveness; Nonny was delusional. At this juncture in our Berkeley soap opera, in the spirit of the emerging sexual mores of America, Santa Rosa mistletoe was added as a new prop on our increasingly cluttered stage. Lary read an article in *The Berkeley Barb*[32] about ancient abortifacients in which the author suggested the use of Santa Rosa mistletoe for this purpose.

Lary and Jim drove up to Santa Rosa in the MG, while I babysat Nonny and Olivia. The two barely-married men returned with a gunnysack full of Santa Rosa mistletoe, their arms and faces covered in scratches.

I do not have any explanation for the next act of collective madness, except the four frontal lobes of our brains were evidently not yet ready for prime time. Lary cut the bottom out of an old chair and we centered a neighbor's hot plate on the floor beneath the missing seat. Filling Nanie's old canning kettle with mistletoe, we covered the mistletoe with water, and placed the kettle with its contents on the hot plate. When steam began to rise through the seat opening, we placed a weeping, bare-assed Nonny on the newly created throne-with-a-hole, draping her lap and legs with a sheet for meaningless modesty.

Nonny sat over the steaming kettle of mistletoe for two days until her buttocks and labia were covered in blisters, her feet and legs swollen into masses of unrecognizable flesh. You could smell the mistletoe all the way across Derby Street. Everything we ate tasted like mistletoe for weeks afterwards—a flavor best described as a complex mix of Mentholatum, cinnamon, and bitter over-cooked Brussel sprouts. The only result was that Nonny lost her appetite. Not a drop of blood was seen, nor a single cramp felt.

Three weeks later, after climbing a high wall of expensive legal obstacles, using lies and madness as her ladder, Nonny obtained a legal abortion in a San Francisco hospital. Then she and Jim dove head first into psychotherapy, both individually and as a couple.

The front row seats Lary and I had to their relational drama helped us gain some ground in our own struggles to climb the mountain of maturity. Lary stopped his active betrayal of our war-induced union, while I felt more confident in my role as Olivia's mother. Every week we went to the Berkeley Public Library for new books. Her favorite was still *Raggedy Ann*. I made her cotton print dresses on Nanie's Singer with deep hems, trimmed in rickrack.

-41-

The Accident

Olivia was three years old when we left Berkeley. Our withdrawal from the confusing cultural chaos was made possible by Norah Brotman, a Berkeley teaching assistant, who offered me the use of her property in Northern Idaho. Norah had one caveat: her house had been vacant for several years and she expected me to make the necessary repairs. Free rent seemed to offer the possibility of stepping off the economic treadmill—the one that kept me moving, while never gaining any miles. My plan was typical of a twenty-something's approach to economic security during that era.

Norah Brotman's property was one hundred sixty acres, coincidentally located on Nora Creek Road,* near Troy, a tiny berg twenty miles due east from Moscow, Idaho in Latah County. Norah told me there was a house, a shed, a pond, and that most of the property was forested. It sounded like a wonderful place.

I made the decision to take Olivia to Idaho in January 1966, eight months before Nonny's pregnancy. I assumed Lary would not be moving with us because he was still involved with Anne. I planned to finish writing about the impact of the Free Speech Movement (FSM) on the governance of UC Berkeley, while providing Olivia with a closer relationship to the natural world.

The move to Idaho required a full-press approach to saving money. Not only did I need funds to buy a used car to transport us to Idaho, I also needed savings for six-months of minimal living expenses. In addition to my full-time job at the library, I found part-time work during school hours administering the Stanford-Binet Intelligence Test in the halls of Oakland public schools. Olivia accompanied me; quietly drawing and watching time wasted, while I measured the IQs of third graders. When the school year ended, I wore roller skates on weekends, working from nine o'clock at night to three o'clock in the morning

---

*There was once a logging camp known as "Nora" near Nora Creek.

as a car hop at a burger joint on south Shattuck Avenue. Whenever Lary was with Anne, Rebecca came to stay with Olivia.

It was late August when Lary told me he had decided to join our Idaho adventure, offering mixed and confusing motives. First, Lary had decided Norah Brotman's place might allow him to shelter to young men trying to slip into Canada to avoid the Vietnam draft. *(I judged this idea emerged from guilt about the special nature of his own exemption.)* Second, in a few weeks Anne would be leaving Berkeley to travel in Europe. *(Was this the real reason he was joining me?)* And third, he had suddenly changed his career goals, deciding to become a luthier, a maker of string instruments, instead of an architect. *(This change was prompted by his having started classical guitar lessons with Anne's guitar teacher and because half our generation, depending on gender, wanted to emulate either Joan Baez or Bob Dylan.)*

Despite his confusing motivations, Lary's decision meant Olivia would not have to face another paternal loss. I felt relieved, indeed, almost grateful—as if I was Debbie Reynolds and he was Eddie Fisher deciding to come back to me, leaving Elizabeth Taylor behind.

In a surprising sacrifice, Lary sold his MG in order to buy a truck to transport us to Idaho, after which Les Carpenter acquired a used 1958 Ford pickup at a telephone company auction with five hundred ninety-five dollars of the proceeds. Lary purchased wood-working tools and rare woods for his new career with the remaining monies.

In mid-October, a month after the mistletoe debacle, we packed most of our books and papers, including my collection of diaries and journals—mailing them to Norah's Idaho address at the inexpensive book rate. At the end of October, we had a yard sale to reduce the size of our household furnishings to what would fit into the bed of the Ford pickup. The Chinese crates and the Emeryville couch sold at ridiculously high prices in the first hour. Just like Ayn Rand's heroic architect, Howard Roark, Lary Carpenter was a man at the cutting edge of his design time.

The night before our departure, Nonny and Jim joined the three of us for a last supper in San Francisco's Chinatown. After dinner we went to see *Dr. Zhivago*. Olivia fell asleep at the sound of the first gunfire. Nonny and I ran out of both hankies and Kleenex, necessarily wiping our noses on sleeves. *Dr. Zhivago* was the first movie I had seen since Olivia's birth. For me, Dr. Yuri Zhivago, as played by Omar Sharif, became the embodiment of a perfect male—

both doctor and poet, he seemed to have equal amounts of testosterone and estrogen. I fell in love with his imaginary persona that night and never let go. Only recently did I realize that Julie Christie's portrayal of Lara Antipova, a young woman who became pregnant out of wedlock, may have been a balm for my own sexual sins.

We three emigrants left Berkeley before the sun was up on the first of November. Olivia sat in the middle, her legs barely reaching the edge of the seat. We had planned a long drive for that day—almost six hundred miles from Berkeley to Ontario, Oregon. For Olivia it was an exciting adventure; she was also looking forward to crossing paths with "Grandpa Frank and Grandma Margaret."

My father, along with my mother, had been attending an entomological conference in Bend, Oregon. After a convoluted planning process, an overnight reunion had been arranged at a motel in Ontario, Oregon, near the Idaho border. The visit would be the first time Olivia had seen her maternal grandparents in a year and the first time my parents would have seen their second-pseudo-son-in-law since the night of my shotgun marriage to Dick. My anxiety had risen as soon as my parents announced plans to stop in Berkeley after the conference. I finally arranged to meet them half-way, my attempt to try to contain contact with my mother at a neutral site. It turned out to be a fateful avoidance.

Heavily loaded, the Ford pickup struggled up the Western side of Donner Pass before descending the Sierras. The two hundred miles between Reno and Winnemucca were desolate, made more so because the recently retired telephone truck had no radio. It was after two o'clock in the afternoon as we pulled into a Winnemucca gas station. The attendant told us there would be little or no traffic on Hwy 95N and he was right. Beyond the rolling foothills, we could see the majestic snow-capped Owyhee Mountains.

About thirty miles after crossing the Oregon border, we made a slow ascent up a hill. At the top, a long stretch of downhill road lay before us; to save fuel Lary let the truck's load carry us down the hill. Only one other vehicle was visible—a slow-moving hay truck, pulling a trailer stacked high with hay bales which was traveling in the same direction. Even though the hay truck was a half-mile ahead, the vocals of a country western song were audible. Lary signaled before moving into the other lane in preparation for passing. As an extra precaution, Lary flashed our headlights and sounded the horn as he gained on the hay truck, approaching the point of passing.

Suddenly, without warning, the driver of the hay truck began to turn left into our path. The driver had not signaled his intention with a hand signal, tail light, or brake light. Nor was there any visible sign of an intersection on the opposite side of the road. The truck seemed to be turning into an open field for no reason. It was too late to stop or slow down. Instead, Lary moved onto the shoulder of the road, pressing hard on the gas pedal in an effort to avoid a collision.

For a few moments, it seemed possible we could escape the looming calamity. The hay truck driver was a teenager, lost in the loud music. Even when I was eye-to-eye with him as we passed, the youthful driver seemed to lack any awareness that another vehicle was on the road and he kept turning the hay truck toward ours. At the very last second, his left front tire clipped the rear right corner of our bumper.

Our Ford pickup rose in the air and began to turn in a half circle. I pulled Olivia behind me in an effort to protect her from the impact. Death felt imminent, filling me with a strange sense, not of fear, but of submission; I began to experience a life review—a time-line of images passing through my visual field.

We three human passengers were flung out the truck's doors and windows. From beneath the tarp covering the truck bed came Spotty and Stripey along with our most essential household goods. The sixteen-year-old driver of the hay truck, testified in court eleven months later that our pickup was airborne almost twenty feet before it landed upside down, crushing the cab into an eighteen-inch stack of metal pancakes. The 1958 Ford pickup had been built long before mandatory seatbelts. Had we been belted, the three of us would have died instantly upon impact. Instead, some angel of accidents lifted all three of us into the air, depositing us in a sloping basin of Oregon mud on the east side of the roadbed.

Lary was thrown out the driver-side door. When he hit the ground, three of his lumbar vertebrae broke and his spleen split almost in half. I was ejected through the passenger side window; the blows to my head from the glass window and ground fractured my skull in two places and cracked a vertebrae in my neck. My face and scalp were badly cut from breaking the window glass.

The force of my body blew open the passenger-side door and Olivia's body followed. Because of her light weight and the absence of obstacles, Olivia traveled fifteen feet past Lary and me. As she slid across the basin face down, her nostrils and mouth were filled with wet Oregon clay. I struggled to crawl to Olivia's side, thinking she had been gravely wounded because she was covered in blood, not realizing the blood was coming from my own wounds.

When the teenage hay truck driver got out of his truck and looked down at the scene from the edge of the road, Lary shouted at him, "Get an ambulance!"

The teenager said nothing. He slowly turned away and moved out of sight. Lary could hear the trailer being detached from the hay truck and a mumbled exchange of words between the driver and his passenger, another teenager. Neither said anything before they sped away. Lary assumed we had been abandoned.

The two teenagers drove north in the direction of Jordan Valley, seventy miles further north on Hwy 95, the nearest place with an ambulance. Jordan Valley had one resident trained as an EMT, but he who was out of town. After the teenage driver described the scene, someone notified the Oregon State Patrol. When the volunteer ambulance driver, accompanied by Jordan Valley's only policeman, arrived at the accident site sixty minutes later, an Oregon State Patrol car was close behind them. The whereabouts of the teenage driver and his passenger were unknown. The only evidence of their previous presence was the double hay trailer on the side of the road.

The nearest trauma hospital was in Caldwell, Idaho—one hundred thirty miles further north. Both the policeman and the state patrolman had some basic first aid training. Wiping away blood and clay, they cleared Olivia's airways trying to make certain that she was getting adequate oxygen.

The ambulance was equipped to carry two patients. After the state patrolman bandaged my head in an effort to slow the bleeding, the volunteer driver and policeman loaded me onto a gurney, while the state patrolman made an assessment of Lary's injuries. After the gurney was loaded into the ambulance, the policeman carried Olivia in his arms, lay her next to me, placing an oxygen mask on her face, and covering us both with an insulated blanket. All three first responders were needed to move Lary, who screamed in pain at their first attempts. After they managed to get a back board and neck brace on Lary, they secured him on the second gurney and somehow managed to load him into the ambulance.

Before shutting the rear door, the state patrolman asked if there was anyone who should be notified. Lary gave him the names of my parents and the Ontario motel where they were waiting. The patrolman relayed that information to the Ontario police. They escorted Frank and Margaret to Caldwell Memorial Hospital, their two-car caravan arrived an hour before our ambulance. For years afterwards, in one of his many repetitive denunciations of governmental waste, Frank deemed the police escort unnecessary.

"We could have found our way to the hospital without their help," he said. "It was a terrible waste of taxpayer services."

My father was a proactive protector of American taxpayers his entire life.

We subsequently learned that the state patrolman took pictures and measurements at the site and arranged for the remains of our pickup to be towed to a junk yard in Jordan Valley. He noted in his written report that the "overloaded hay trailer had no license or tail lights."

Within an hour of our arrival at Caldwell Memorial Hospital, both Lary and I underwent surgery. Lary's spleen had to be removed. Bone cement and a plate were used to stabilize his fractured vertebrae. Parts of my skull were cut away to allow for swelling; my facial lacerations were sutured and I was outfitted with a neck brace after x-rays revealed my third cervical vertebrae was cracked.

Miraculously, not a single scratch or bruise was found on Olivia's body. While she had no visible soft tissue wounds or broken bones, pneumonia had taken hold from the mud she had aspirated and the long period of exposure. The three of us spent the next several weeks in Caldwell Memorial Hospital. Olivia shared a room with me, while Lary was assigned a bed in the Men's Ward.

During those first critical days after the accident, my parents were essential helpmates. At first light, my father drove to the accident site—easy to find because it was still covered with our scattered belongings. He went in search of Olivia's two kittens, who had been thrown from the truck with the same force as the human passengers, as well as her blankie. Olivia had repeatedly asked for "my blankie" and "my Stripey and my Spotty" throughout that terrible long night in the emergency room.

My father found no sign of either Stripey or Spotty. The door of their carrier was missing—only a clump of Spotty's hair on a half-hinge gave evidence of their prior existence. He set out both water and food by the kennel. Always an expert finder of lost items, my father made a thorough search of the area, while simultaneously gathering household items, clothing, and Lary's tools. Olivia's blankie was nowhere to be found.

My father wondered whether the Olivia's blanket might still be in the cab of the truck. He drove back to Jordan Valley and located the junk-yard, weeping at the site of the crushed cab. The helpful proprietor used a jaw-like tool to create access to what had been the interior of the cab and Olivia's blankie was pulled free, dripping gasoline. My father placed it in a plastic garbage bag in the trunk, next to the muddied assortment of items gathered from the site. He was still teary as he drove slowly back towards Caldwell.

At a laundromat near the hospital, my father spent the evening washing the salvaged clothes and linens from the accident site. Olivia's blankie required bleach and four wash repetitions because not only was it saturated with gasoline, it was also stained with what appeared to be axle grease. It was after eleven at night when my father returned to the hospital. Olivia was sleeping. His daughter and second son-in-law were both still in the ICU. Tucking Olivia's blankie under his granddaughter's arm, he woke Margaret from an uncomfortable nap in a chair. The night nurses encouraged the couple to get some rest. Exhausted, they drove to a nearby motel room and collapsed.

The next morning when my father visited Olivia, she said, "Thank you, Grandpa, but my blankie is not the same."

All the familiar scents of security had been first torn from her, then washed away.

Frank made the five-hour, four-hundred-mile round-trip to the accident site in search of Stripey and Spotty each of the next three days, leaving fresh food and water and searching in potential places of shelter. He salvaged any recognizable household items, while also gathering the destroyed remnants for proper disposal in the motel dumpster. On my father's last trek, a coyote was waiting for him by the empty feline food and water bowls.

My mother began to search for temporary housing for us after Dr. George R. Wolf, the physician in charge of our care, explained we would be discharged on different dates. My mother located a small furnished attic apartment in the home of a Mormon widow. Told of the circumstances, the widow agreed to a short-term rental of fifty-six dollars a month.

When all hope was lost for Spotty and Stripey, and the site of the accident had been transformed, leaving no trace, my father wrote a check for a 1952 Chevrolet pickup. He acquired the necessary Idaho license plates, and had his

State Farm agent in Tucson prepare a comprehensive insurance policy, sending a check to cover the first year of premiums. Lary had purchased only liability insurance on the Ford pickup. My father knew that the hospital and ambulance bills were certain to wipe out my savings for our Idaho living expenses. At the end of the second week, my parents resumed their journey home. Both were exhausted from their supportive efforts and my father had pressing work at his Tucson research lab.

I was discharged five days after my parents left Caldwell. Following my father's careful directions, I drove to the grocery store which he had determined had the best prices. With a bag of essentials, I drove to the home of the Mormon widow to pick up the key and receive the requisite tenant instructions.

That night, I sat down next to Olivia's hospital bed to review the charges up until my discharge. The sub-total already exceeded my savings. By then we knew that neither Keith Taylor, teenage driver of the hay truck driver, nor his father, the owner of the truck, had insurance. The next day I went shopping for a lawyer.

Caldwell Attorney Peter J. Boyd, a partner in the law firm Gigray & Boyd, agreed to take our case on a contingency basis. He took photographs of my face and my skull, explaining that by the time the case went to trial some of my obvious scars would have faded. Boyd was certain there would be a trial because he knew Don Taylor, the man whose son had been driving the truck, who had already given a statement to a local police reporter.

"The accident was not my son's fault," he reportedly said, "but was caused by reckless California hippies who came to squat on Idaho acreage."

Olivia was discharged the following week, although she had to return for outpatient respiratory therapy on a Nebulizer machine twice each day for a fortnight. Lary wasn't able to join Olivia and me in the temporary attic apartment until the fifth of December. His first efforts to climb the interior stairs to our attic door revealed the severity of his injuries and the debilitating effect of his long confinement.

As a recovering threesome, we groped our way through the holidays. I helped Olivia decorate a small tree using traditional supplies of popcorn, cranberries, and colored paper. Our sympathetic Mormon landlady lent us a string of colored lights.

On Christmas morning the Mormon widow left wrapped presents for each of us outside our door: scarfs she had knitted for Lary and me, along with two hot pads. The widow had gone overboard for Olivia, knitting her a colorful poncho with a stocking hat. At the bottom of the box, beneath the poncho, Olivia discovered a small doll dressed in a matching outfit.

Watching Olivia's delight, I began weeping. The three of us then descended the stairs to thank the widow for her kindness and generosity—for introducing Olivia to the spirit of Christmas.

-42-

Starting Over

We resumed our journey to Nora Creek Road on New Year's Day 1967. Olivia began asking if we were "almost there" before we even reached the outskirts of Weiser. She was the most anxious of our threesome, but it wasn't easy for any of us to resume our pre-accident positions in a truck cab. Pulling Olivia close for comfort, I was grateful for the distracting sounds of unfamiliar music and agricultural reports coming from the Chevrolet radio.

Our few remaining belongings, some essential foodstuffs, six bundles of firewood, and twenty-eight gallons of water were in the truck bed. After passing through Weiser, we were once again in under-inhabited territory. I made Lary promise that there would be absolutely no passing. When we stopped for gas at the junction of US Route 95 and Idaho 55, we ate the baloney sandwiches provided by our landlady. There seemed to be few other vehicles on the road; it was not until we reached Grangeville that we experienced what passed for traffic in rural Idaho, a three-way stop where two pickups waited.

The homemade sign marking the turn-off to Nora Creek Road was so small, we missed the turn. Norah had drawn a map and shown us photographs of her house. It was located about six and one-half miles north of the juncture with Hwy 8. The first structure we passed was Bendel's Meat Packing Plant, before we entered a heavily wooded area where tall cedars mixed with a few white pines stood crowded together on either side of the road. The single lane road was narrow, unpaved, deeply rutted, snow-packed, and icy.

Olivia was the first to spot the house as we came around a slight bend in the road and the forest gave way to open fields of snow on either side. One third of Norah's property had once been cleared for wheat. Those fields lay fallow, having been put into our nation's Soil Bank; the remaining acreage was dense, old-growth cedar forest. Her house was situated near the top of a gently sloped five-acre field, shaded by several large cottonwood trees.

In architectural lingo, the house was a "salt box"— the side walls covered with wide strips of grey asphalt shingles, the steeply pitched roof in cedar shakes. A covered porch sheltered the single entrance on the west side of the house. Below the house, closer to the road, a small ancient barn tilted away from the slope. Norah's hand-drawn map showed a well located just above the bed of Nora Creek. Lary drove up the slope closer to the house and turned off the motor. We had finally arrived at our new home.

The House on Nora Creek Road, January 1967

No human had lived in Norah Brotman's house for several years; consequently a large number of spiders and mice had sub-divided it into *Arachnidic* and *Rodentia Murinae* condos. A wood-burning stove sat in the center of the ground floor. To its left, against the stair wall, was an ancient upright piano, fully occupied by hairless baby mice, whose progenitors had eaten the felt key pads. The kitchen area was on the right side of the room, the boundary marked by an empty, mold-infested refrigerator.

Norah said a deposit had to be made with the local electrical power cooperative to restore service and we had been warned that the pump on the well might be broken. Having arrived at our new home in the middle of winter, we found it crowded with the nests of others; without electricity or running water, and the only source of heat was the wood-burning stove.

Lary got a fire going with a bundle of wood from Caldwell, while I wiped the dust and spiders out of Norah's cast iron tea kettle, rinsing, then filling it with water from one of the gallon containers we had brought. I made tea for Lary and me and Ovaltine for Olivia. We sat huddled around the wood stove on two half-broken kitchen chairs and a sturdy crate, while I, in the tradition of Afton

Evans, attempted to make some Wonder Bread peanut butter sandwiches on my lap, easily the most sanitary surface.

We ate, sipping tea and Ovaltine, absorbing the warmth of the stove, trying to ignore the sounds of the other tenants. After cleaning our faces and teeth as best we could, we awkwardly made separate use of the portable commode found on the porch. Lary unrolled our sleeping bags, laying them out on a stack of wool army blankets my father had thoughtfully acquired for us from the Army Surplus Store in Boise. We kept the fire going all night, while making a cocoon around Olivia, who actually slept quite deeply; the two adults, not so much.

In the days that followed, long hours of hard work produced a steady stream of small improvements in the habitability and charm of the house on Nora Creek Road. At night, when my aching muscles prevented sleep, I came to see how badly I had underestimated the psychological and physical adjustments required for this move, multiplied and squared by the accident. My life lesson: There is no such thing as free rent.

I was slow to understand why Olivia seemed to be so happy, obviously savoring our adventure in the Idaho woods. My presence at her side day and night was something she had never known. Olivia was now three years old, a developmental moment when the absence of distance between us was optimal.

House on Nora Creek Road, with porch protected by fire wood, March 1967

-43-

Cross-Species Relationships

As if we were actors in *Dr. Zhivago* escaping to an ice-covered house in the country, we wore woolen gloves day and night. The windows were frosted over, turning the wood-burning stove into a human magnet. We headed to the local lumber mill for a load of slag to keep the stove going until we could cut our own wood. Before leaving Berkeley, Lary had estimated we would need ten cords of wood to get through the winter.

Glenn Gilder was the first neighbor to extend a hand of friendship to us. At age sixty-five, he was known simply as "Gilder" by almost everyone in the Troy area. Passing Norah's house on his way home from work, Gilder stopped to find out if we were illegal squatters. Our candlelight, chimney smoke, and unknown truck made him appropriately suspicious.

Laughing at my jokes through multiple broken teeth, Gilder's long face was craggy, his barn coat well past its expiration date, and his cowboy hat stained by a band of sweat. After that first encounter, Gilder often conjured up a need to drive by the house to watch us, sometimes with amusement and sometimes with admiration, as we struggled to survive.

Two weeks after our arrival, the pump was delivering well water to the kitchen sink, although we never drank anything but bottled water after our close-up examination of the well. While Lary worked to put Norah's ancient chain saw in working order, I concentrated on deep cleaning and systematic spider and rodent control. With scrap lumber from the ancient barn, we built a four-foot square sled-like device, attaching a piece of sheet metal to the bottom in lieu of runners. The sled was so heavy it took both of us to pull it by rope to the top of the hill at the edge of the forest on the opposite side of Nora Creek Road.

Approaching the forest with awe and artful intentions, Lary used the chain saw on the dead wood on the forest floor, before we worked to bring down the weakest trees where more sunlight was needed. After I used a handsaw to strip the trunks clean of branches and boughs, we cut the trees into thirty inch sections, the length of Norah's stove box. Olivia became an expert hunter of pitch wood, a sticky, resinous substance found on conifer trees used as fire starter, religiously filling an old canvas backpack each day. Pitch was essential for fire building in the absence of daily newspapers.

Once our hand-made sled was loaded, the weight of the wood took the driverless-sled down the hill until it slid onto Nora Creek, then frozen solid. From there we loaded the wood into the back of the pickup before driving it close to the back porch to be stacked into cords that created a wind break. During that first year, we were rarely more than a day ahead of the stove in our splitting, a skill at which both Lary and I were soon proficient.

In mid-February, the Chevrolet pickup truck broke down after dark in the middle of a snowstorm when we were halfway between Troy and the turn off to Nora Creek Road. While Lary opened the hood and began tinkering in the dark, Olivia and I walked to the nearest farm in search of help. As we approached, a young farmer came out of the barn, telling us he had a wife and two children in the house, but no telephone. Taking immediate control, he told Olivia and me to wait in the barn before picking up his toolbox and making his way to Lary and the unresponsive Chevrolet.

Inside the barn, Olivia spotted a litter of five kittens. I watched as she fell in love with the only calico, whose most predominant color was grey, accented by patches of white and apricot. The calico's face had an unusual open sweetness. While we sat on the barn floor using sprigs of hay to tease the kittens into play, the young farmer helped Lary clean the spark plugs, which proved sufficient to restart the motor. When the farmer came back into the barn, he quickly made an assessment of Olivia's wants and my vulnerabilities.

Using normal Northern Idaho vernacular, he turned to Olivia and said, "Say, young lady, why don't you take some of those kittens home with you? I gotta drown 'em in the water trough in the morning. We got too many!"

As soon as the farmer spoke, I knew we were about to acquire replacements for Spotty and Stripey. Olivia selected the friendly female calico and a male tabby with a coat similar to that of Norrie and Stripey. In lieu of a tail, he had a short rabbit-like stub. Riding home, Olivia began addressing the two new members of our household as "Grey Cat" and "Tiger Cat" while they happily snuggled in her lap.

A month later, the three of us, along with Grey Cat and Tiger Cat, were jolted awake in our attic beds by an invader. At the sound of the hinges being ripped from the exterior door frame, Lary started barricading access to the upstairs by dropping Olivia's new mattress to the bottom of the stairwell, bracing it against the thin paneled door with a trunk full of abandoned Brotman family clothing.

Our uninvited visitor was a large bear. We listened as the bear snorted his way through the food in our kitchen. Bear novices, we tore up our top sheet and tied the pieces into a make-shift knot rope in case we had to escape through the single attic window. The rest of the escape plan was best not considered. The keys to the truck were on the kitchen table and it was unlikely we could outrun the bear.

At dawn the house was quiet. From our attic window, we watched the lumbering ursine silhouette of a large male bear with visible gonads disappear in the dark shadows at the edge of the forest. Lary removed our improvised barricade and we timidly descended the stairs to assess the damage. Our newly restored nest was a mess, reeking of the bear's strong distinctive scent. The bear had eaten all six loaves of our homemade bread, two jars of honey and peanut butter, and the three pints of homemade choke-cherry jam Gilder's wife Agnes had given us the previous day. The lids of the canning jars had been removed with apparent ease, evidently by using a single claw. Cleaning up, I remembered the mother bear in Yellowstone who loved my mother's cherry chocolates.

Lary and I lacked the funds, had no shooting skills, and were against the ideology required for the possession of a gun. Yet, we were living in the middle of the forest, without a telephone, or even near neighbors. We had come to the wilderness to avoid the use of weapons, even for our own defense.

We decided by unanimous vote to add a canine to our household. Both Grey Cat and Tiger Cat were allowed to vote on this addition. Olivia was very much into taking polls and counting hands, as if she had learned the democratic process while watching Berkeley's Free Speech Movement from her stroller.

When the bear mess was under control, we drove into Troy for fresh food supplies, door hinges, and framing wood. Inside the Troy Market, Lary saw a handwritten sales notice on the bulletin board for Australian shepherd dogs in nearby Deary. The market was the touch point for everything. Troy had been settled by Swedish Lutherans and the owner still advertised his weekly specials in Swedish. We solidified our reputation as stupid foreigners during our first shopping expedition by asking the meaning of "*lok pa rea*" (onions on sale) and "*potatis till forsaljning*" (potatoes on sale). We bought our second quart jars of peanut butter and honey that week, paying with cash to quizzical looks.

Back home, I started a fresh batch of bread, while Lary fixed the door. Olivia kept busy trying to draw a picture of a bear—the previous night's visitor was very much on her mind. Late in the afternoon, we drove east to Deary and on our third try located the farmer who had posted the notice.

Unusually adamant about gender, Olivia kept repeating over and over, "I want a boy dog because he will protect us better!"

Born with an unusual ability to survey the world, Olivia had decided we two females would never have made it alone on Nora Creek Road.

The old farmer, exhaling what smelled like pure grain alcohol, led us to a shed near his sheep pens. We paid thirty dollars for the first of three almost full-grown Australian shepherd pups who moved toward Olivia, naming him Pablo after Pablo Picasso because his rear legs seemed to be similarly bowed. His coat was silky, mostly black with strategic touches of auburn and white.

At eight months, Pablo quickly demonstrated the high intelligence of his breed and soon chose Lary as his alpha human. We had no indoor toilet that spring, so whenever Lary went out to relieve himself, Pablo did the same. His large brown eyes watched as Lary lay down on the floor by the wood stove, observing that Lary allowed Grey Cat and Tiger Cat to climb onto his body for warmth and comfort.

Tiger Cat and Pablo with Lary, Nora Creek Road, May 1967

Within days Pablo was giving Grey Cat and Tiger Cat the same prerogative, and both cats quickly came to prefer Pablo's body to Lary's because it was softer, warmer, and stayed still longer. After Pablo came to live with us, the only sign we saw of bear was occasional scat near the shed.

Lary created one of the best swings in the world for Olivia. He hung it in a huge cottonwood tree that grew just above the house, crafting a comfortable seat of cedar with buttock-like curvature, sanding it as smooth as a guitar. Lary also taught Pablo to take hold of a rope tied to the seat. At Lary's command, "Pull, Pablo!" Pablo would grab the rope tail hanging from the seat of the swing, run up the hill, pulling Olivia on the seat until she was positioned at the highest point. Pablo held the rope tight in his teeth until Lary gave the command "Release!" Then Olivia would soar, pigtails flying behind her. She was fearless and ecstatic on that swing. If Olivia remembered to pump her legs, she was in motion for as long as she wanted. In this respect, Pablo was truly a working dog, since, as a canine rope-puller, Pablo made it possible for Lary to work without interruption.

Olivia on swing with Pablo, October 1967

When the ground started to thaw, I started to prepare a garden plot. Gilder drove by while I was attempting to spade the thirty-year-old sod that resembled cement by jumping on a shovel with both feet in boots with soles so full of holes, the addition of cardboard was a daily necessity. As was Gilder's way, he just slowed down, smiled, and waved.

247

Pablo with Lary by the pond on Nora Creek Road, October 1967

Early the next morning Gilder returned with his tractor and plow. In three hours he created two huge garden plots, one across the road at the top of the hill by the pond and one in the vale just below and south of the house. Gilder thought the hill garden would be more vulnerable to hungry deer and the vale garden more susceptible to frost, but between the two gardens we would be able to grow enough to meet our needs. The following morning, Gilder returned with a trailer of seasoned manure from his barnyard and two large gunny sacks full of potato (*potatis*) and onion (*lok*) starts.

After Gilder plowed our gardens, we invited him to share a meal with us, the first dance in an important social exchange. The second was when his wife Agnes invited us for Sunday supper at their place. The meal trades went on, but Agnes only came to our house once on Christmas Day 1967. She was fearful of Jews; when she learned my maiden surname, the friendship took on a permanent distance. I never learned what she felt about other ancestors in my bloodline. If her cooking hadn't been so incredibly good, I would have honored her prejudices by letting Lary and Olivia go alone to sit at her table, but on Nora Creek Road hard labor made me hungry almost all of the time.

During one of those Sunday suppers, Gilder invited us to help him tear down a barn north of Deary because he knew Lary was looking for weathered lumber and Gilder wanted the frame. Lary planned to use the weathered wood from the barn's exterior to insulate the interior walls of the main floor of Norah's house, to build a wall of bookshelves, and to create a half-wall work counter between the dining and kitchen areas. Lary was drawn to aged wood, the kind that had been darkened by weather with swirling knot holes and random spots of vivid orange and yellow moss.

What Gilder did not tell us was that he had shared Lary's outside-inside-barn-wood plan with other local residents over coffee at Troy's Inland Hotel. The use of old barn wood for interior walls was one of the first of many California counter cultural behaviors that made us laughingstocks in Troy.

We three humans and our three animals survived on my meager money that first year because of Bill Bendel's Meat Packing Plant. We could acquire a whole cow liver at Bendel's for twenty-five cents. It took our household a week to eat one bovine liver, including Grey Cat, Tiger Cat, and Pablo. In addition to our consumption of inordinate quantities of honey and peanut butter, the Troy cashiers were gossiping about our consumption of "*lok.*" The only way Olivia and I could eat liver was if it was completely buried beneath a disguising heap of sautéed onions.

Bendel's Meat Packing Plant put us in closer touch with the often unbearably painful links between men and animals, but it was not until Gilder decided to slaughter his sow, Millie, that I fully understood the sacred strands between species. Millie, six years of age, was the smartest in Gilder's eclectic collection of animals. Gilder had a dog, three cows, two goats, several dozen sheep, two working horses, chickens, geese, barnyard cats, and Millie. Millie greeted guests at the front yard gate and slept on the back screened porch to be close to Mrs. Gilder's kitchen scraps. Whenever Gilder was outside, Millie was at his side, more like a dog than his own.

But Millie had been raised to provide sustenance to Gilder's family and the time had come for her to do so. I don't know why we agreed to attend Millie's demise except that it would have seemed California-cowardly to have refused Gilder's invitation.

Also, Lary had decided he needed to learn how to do everything in preparation for the end of the world. Not long after we had settled on Nora Creek Road, Lary picked up the end-of-the-world virus, which was then running rampant among certain residents of Northern Idaho, known as "survivalists." Coeur d'Alene, one hundred twenty-five miles north of Nora Creek Road, had the largest number of acute cases.

Early on a Saturday morning, we drove to Gilder's place. As always, Pablo rode standing up in the truck bed, his body leaning forward toward the driver's cab window, his nose in the wind as if he were trying to be cheek-to-cheek with Lary. When we arrived, I insisted Olivia stay in the kitchen to help Mrs. Gilder with her endless cooking, while Lary, Pablo and I followed Gilder out to the barnyard.

I will die with the visceral memory of Millie's face and the sound she made as Gilder approached her with a huge knife. She let loose a mournful cry, but did not turn away from her master. Then a large amount of blood flowed from her throat. Right after that, Gilder stuck a huge stick in Millie's mouth, ramming it out the back of her head so as to be able to hang her body up for the remainder of the dismemberment. Running as fast as I could to our truck, I crouched behind it. When I could stand up without retching, I got Olivia into the truck and drove home.

Hours later, Lary and Pablo walked home in the dark having spent the day helping Gilder through the entire dismemberment process. Pablo did not eat his evening meal and refused all offerings the following day as well. After a few days of curing, Gilder brought us a pork roast and a side of bacon. I prepared the roast, but Pablo would have none of the offered scraps. From the day of Millie's demise until the day of his own, Pablo ate only dry kibble or sweet potato peelings.

Shortly after Millie's death, I noticed that Grey Cat was pregnant. Olivia and I created a nest for her in a secret niche Lary had crafted between two rafters in Olivia's bedroom. It seemed to be the safest place for Grey Cat and her first litter of kittens, since Tiger Cat might succumb to his natural instincts by attempting to eliminate any newborn male competition. The door that kept the bear out could also keep Tiger Cat downstairs. Olivia carefully cut a strip off an old army blanket to make the nest soft.

The pain of Millie's slaughter was partially healed by the birth of Grey Cat's first litter. She went into labor while I was sitting on what resembled a park-bench, which had been built by Lary to perpetuate the couch drought. I attempted to make it comfortable with a homemade velvet cushion, filled with wool shorn from Gilder's sheep, but the bench was so short only Olivia could take a nap on it. I was sitting on it when Grey Cat jumped into my lap. She seemed to be in search of loving, but she immediately began twisting her body, making strange sounds. After several minutes, a kitten emerged from her body in its birth sac. Grey Cat remained on my lap as she gave birth to four kittens. I was filled with awe as she licked the sacks away and brought her kittens to life. Then I wept with joy at the trust she placed in me. It was as if the spirit of Snowball had come back to me twenty years after her abandonment.

The animals of Nora Creek Road touched the deepest parts of my soul. The devotion of Pablo, the intelligence of Millie, and the instinctual trust of Grey Cat are why cross species relationships remain my greatest source of hope for our world. Three different species shrank distance and differences. When a cat lies down with a dog, when a pig is devoted to an old man, when a feline mother trusts a human mother, they are showing us how to achieve *optimal distance* and mend our broken world.

Olivia and Pablo with the author, Nora Creek Road, near Troy, Idaho, November 1967

-44-

Twin Falls

Spring sprung a special surprise—a field of daffodils came to life around the house, creating a magical botanical setting, similar to the way *Varykino* looked in the film version of *Dr. Zhivago*. A half-century earlier a Swedish farmer's wife must have planted a dozen bulbs, which had been stirred into the soil and spread year after year as the field was plowed.

During the first four months of living on Nora Creek Road, Olivia had been the only child, though never lonely. But she needed friends her own age so I took an unplanned social risk in an effort to end her isolation from other children. I began following a young mother with three young children on Moscow's Main Street. By both clothing and accents, the mother and her children had the attributes of outsiders.

Suddenly, the mother turned around and asked, "Is there something you want?"

Caught off guard by her directness, I blurted, "Do you believe in Summerhill?"

Pat Sullivan laughed in response. The wife of John M. Sullivan, a graduate student in political science at the University of Idaho, Pat knew enough about A.S. Neil's Summerhill School in Leiston, Suffolk, England[33] to respond to my question with a definitive "Yes, although there is nothing like Summerhill around here."

Pat and her children were fun and friendly. After a short visit on Main Street, I invited the Sullivan family to visit us on Nora Creek Road the following Saturday. The Sullivan family brought along friends, another graduate student couple, who had a daughter Olivia's age. Initially shy, Olivia watched the four visiting children line up to take turns on her swing, but she soon began leading the pack.

Olivia introduced them to the bright orange and black salamanders hiding in the woodpile before leading them to a bed of cedar-scented moss at the edge of the forest, not far from a bubbling spring guarded by a large toad. She revealed the hiding place of a gigantic bull snake, the numerous tadpoles swimming in the pond, and the secret entrance to a badger's house beneath the roots of an ancient cedar.

Olivia and Grey Cat on Nora Creek Road, April 1967

After that first visit, most Saturdays became a regular day of play and exploration for the five children and intense political conversations for their parents. The 1968 Presidential election was just beginning to take shape with rumors that Democratic Senator Eugene McCarthy would challenge incumbent President Lyndon B. Johnson on an anti-war platform. In those days, in Northern Idaho, we three couples felt as if we were the only Democrats in Latah County.

Norah's house became habitable just as we ran out of money. After Gilder told Lary there was carpentry work at a granary in nearby Kendrick, Lary drove over on Friday afternoon. The foreman told him to report for work Monday morning.

Leaving a bewildered Pablo behind, Lary left the house early, carrying a sack lunch. The foreman greeted him gruffly with a set of orientation instructions and sternly instructed Lary not to use the cross beams to reach the opposite side of the granary because the beams were too slippery with grain oil.

Lary working as a Luthier on Nora Creek Road, April 1967

Around two o'clock in the afternoon, Lary did what the foreman had specifically instructed him not to do. Instead of climbing down the ladder to the floor and crossing to the ladder on the opposite side, he tried to shorten the transit by walking on one of the cross beams. Lary slipped and fell thirty feet to the granary floor. Were it not for the cushioning effect of the grain, he would have lost his life. Lary survived, but his left leg was shattered and he re-injured his previously broken back.

It was almost dark when Pablo announced a stranger was approaching. The Latah County Sheriff arrived with the bad news. He stood in the doorway and spoke while surveying the interior without looking directly at me.

"Your husband has been in a serious accident," he said, his affect flat. "I've come to take you to the hospital in Moscow."

Quickly gathering our things, Olivia and I left an even more bewildered Pablo behind with the Grey Cat and Tiger Cat and their kittens and climbed into the sheriff's truck.

At the hospital, the doctor used an x-ray to show me how Lary's left leg was broken in five places, explaining, "Your husband faces a long period of recovery and multiple surgeries, the first of which I plan to undertake early tomorrow morning."

Bookshelves and cupboards made of weathered barn wood, April 1967

Olivia and I spent the night in Lary's hospital room because we had no way to get home. Gilder heard about the catastrophe the next morning over coffee at the Inland Hotel, where Lary's fall was already a source of gossip, accompanied by laughter about "the stupid California hippie."

Leaving behind his half-finished coffee, Gilder immediately came to the hospital. Later he drove Olivia and me to Kendrick to retrieve the Chevrolet so we could come and go from Lary's new bedside as needed.

When Olivia and I opened the door to the house, Pablo moaned in relief and consternation; Olivia thought Pablo sounded like he was asking, "Thank God, you're here! But where is Lary?"

The next day I was relieved to learn that even though Lary had been on the job only a few hours, he was miraculously covered by Idaho's Workmen's Compensation Hospital Insurance. Otherwise, I might have been financially forced to let him die. Lary seemed to be incapable of being supervised or subordinate. In today's psychological parlance, he would likely have been diagnosed with some kind of "oppositional defiant disorder." In Lary's defense, Olivia had her ways of reminding me to be grateful for his work ethic. As long as no one was paying or supervising him, Lary worked much harder than most men.

Over the next several weeks Lary underwent three more surgeries. One day while he was still confined to his hospital bed, I left Olivia in his room and went to work as a waitress at the Inland Hotel for Norman Berg. In retrospect, Berg probably hired me hoping my presence would increase his customer base, since most of the locals were curious about the California hippies on Nora Creek Road. The pay was sixty cents an hour plus tips.

At the end of the first long day I had earned a total of six dollars and twenty cents. Since I was already a laughingstock, I proposed a trade. Instead of Berg writing me a check, he could have my day of labor in exchange for an old metal ice cream chair in the back of the kitchen. Covered in grease, it had no seat. Berg thought he had made himself a real deal. I also told Berg I wouldn't be back the next day because I had to find a job that paid a living wage. I still have the chair and feel a sturdy satisfaction whenever I sit upon it.

When Lary was finally discharged from the hospital, I used Olivia's mattress to make a bed for Lary on the main floor, promising Olivia that she could sleep with me. She thought it was a great trade. Sadly, however, I was already scheduled to start work on the only night shift work I had been able to find— as a nurse's aide at the Latah County Nursing Home in Moscow, where my wage was one dollar and ten cents per hour. Since I was a married woman, I was assigned to the Men's Ward where I spent long nights lifting and cleaning between the flaccid buttocks of old men who had lost control of their bowels— a precursor of tasks to come.

Life had been as defiant and dangerous in Tucson as on Nora Creek Road. Shortly before Lary's fall, my father sent a letter to explain that he had to retire.

"Retirement is the only way I can continue to manage Margaret," he wrote.

My mother's disappearances were increasingly frequent. She had put locks on every cabinet door, trunk, or other storage container in the house. She had also cut the telephone cord almost a dozen times, making it impossible for personnel at my father's lab to reach him. Because my mother now believed the FBI was trying to poison her, she was largely subsisting on saltine crackers and apple juice.

My father was still incapable of hospitalizing her. I offered to come to Tucson to handle her commitment, but he declined my offer. Her behavior had worn him out; but he was a committed atheist with respect to psychiatric confinement. I felt sick at heart that his professional work would no longer be a source of distraction or distance from my mother's mind. Although he was only fifty-five, my father was eligible for federal retirement because he had begun to work at the USDA entomology lab in Salt Lake City while still in high school. Frank V. Lieberman had more than the requisite thirty-five years of service.

As soon as I learned my father was going to retire, I conjured up a twin travel plan for my parents. I thought my father would enjoy Europe and traveling would provide my mother with her treatment remedy of choice—constant relocation. My father was interested; Margaret was not. Nonetheless, somehow he persuaded her to plan a European trip.

The day before my parents were scheduled to fly to Europe, my mother stepped off the curb in front of their Tucson home, stumbled and fell, breaking her left leg in three places. Her accident followed Lary's by only a few days, but a new postal employee in Troy kept returning my father's letters and all our other mail stamped with the notation "Addressee Unknown."

My mother's recovery was slow. The bones in her leg had to be screwed back together; mistakes and infections necessitated multiple surgeries. The only upside of her fluke accident was that her surgeon kept her on Thorazine during her serial hospitalizations.

-45-

The Company That Came

When Lary's back and leg were healed enough for him to walk with a single crutch, his parents, Les and Betty Carpenter, came to see their twice injured son and his "excuse for a family"—the phrase Betty used to describe Olivia and me within our earshot.

An inexperienced and pre-exhausted hostess, I was ill-equipped for company. We had learned to eat within our budget, but Betty and Les were not accustomed to a diet of liver and onions with chard or peanut butter and honey sandwiches with wild lettuce. I emptied our coffers trying to feed them more typical fare, trying to win acceptance from Betty—a futile effort.

After the Carpenter's ten-day visit, Nonny and Jim arrived. I was working nights at the Latah County Nursing Home and most of my meager salary was going to cover Lary's uncovered medical expenses and our required payments to Caldwell Memorial Hospital. My sleep period that summer ran from five thirty in the morning, when I returned from the Latah County Nursing Home, until about seven in the morning, when Olivia awoke. I was sleep deprived and undoubtedly sometimes irritable from exhaustion during their visit, but Nonny seemed unusually distant and withdrawn. She and Jim were both still in individual and couples psychotherapy, but had dropped out of school and were working for the Glide Foundation[34] in San Francisco while living in San Rafael.

They were intensely evangelical about the value of therapy, repeatedly questioning Lary and me about our self-awareness and life goals. Although we agreed with their judgment that we were somewhat lost and accident prone, Nonny and Jim found our plans inadequate and Olivia's chore routine (pet feeding, table setting, and bed making) too heavy a burden for a child.

When my mother was finally able to walk with a cane, she insisted on selling their red brick house because she was convinced her orthopedic surgeon was trying to kill her. Further, she wasn't safe in the house because the mail containing bills from her surgeon provided proof he knew where to find her.

Their Tucson house sold quickly because of my father's obsessive maintenance. Having put their household furnishings in storage, on the day of the closing they climbed into their new 1967 Pontiac Catalina sedan, and began driving to Idaho, pulling a small U-Haul trailer. The contents of the U-Haul included the comforting green platform rocker of my childhood and my father's professional library. The rocker was for Olivia; my father's library was to be donated to Washington State University (WSU) in Pullman.

My parents were our third set of company; they stayed with us for two weeks before departing on their European tour. As we had for Les and Betty, as well as for Nonny and Jim, we gave my parents the upstairs bedroom. Olivia and I used our sleeping bags on the main floor, where Lary was still sleeping on Olivia's mattress. This period of co-habitation was harder on my mother than me, but it was very good for Olivia. She built a solid relationship with her Grandpa Frank, who loved exploring the outdoors as much as his granddaughter.

Olivia and Grandpa Frank hiked through the surrounding woods and fields, searched for the headwaters of Nora Creek, and made a book about the birds they had seen and identified. The pond was thoroughly explored and they created another book detailing the creatures large and small residing in our woodpile. My father also took charge of providing meticulous care of our two garden plots, while giving Olivia lessons in horticulture. In charge of the daily harvest, the pair took equal pride in preparing vegetables for our meals.

My parents flew from Spokane to Chicago, where they took a connecting flight to Frankfurt, Germany, leaving their new Pontiac sedan in our care. Watching the two of them cross the tarmac and climb the ramp stairs, my mother slowed by her cane, I began to weep. I felt both relief and fear—relief that our crowded house could return to normal occupancy, but fear for my father. At the very bottom of my long list of worries was my failure to persuade my mother not to wear her ubiquitous panty girdle on the long flight across the Atlantic Ocean. Driving back across the Idaho border, I realized the girdle was probably the least of my mother's discomforts.

Upon landing in Frankfort, Germany, my parents took delivery of a new Beetle, an understandable choice for an entomologist. My father wrote me one airmail letter every week. In between his reports on the price of tomatoes in the various countries, I could tell he had his hands full. My mother sent only cryptic postcards, believing all her communications were being read by the FBI.

Before they left, I had obtained a position as a teaching assistant in the language department at Washington State University (WSU) in Pullman, just across the Idaho-Washington border from Moscow. I resigned from my "married woman" duties at the Latah County Nursing Home. My WSU salary provided enough income that we enrolled at the University of Idaho (UI). I had two requirements to finish and Lary had decided to return to his Ayn Rand plan to finish his architecture degree. Between the accident, the hard work of making Norah's house habitable, and his work-place injuries, his career as a luthier never took hold.

As in Berkeley, I worked and went to school, while Lary went to school and maintained the house. We found a sunny daycare center for Olivia located in a Methodist Church in Moscow. The hours she spent there gave her contact with other children, as well as an independent space. On Mondays, Wednesdays, and Fridays, I would drive eight miles from the UI to WSU. Not only was my WSU salary more than worth the trip, I loved driving the Pontiac. Apparently my body was still producing car hormones.

On September 29, 1967, I placed a long distance call to Nonny from a telephone booth on campus. I had not heard from her since she waved goodbye on Nora Creek Road, and I had had several dreams that indicated something was wrong. When Nonny picked up the phone, my best friend seemed even more distant, her cool demeanor barely recognizable. The call was short, but she promised to write me a letter of explanation. When it arrived, Nonny's letter was a devastating personal critique of me and our friendship.

> "It is so difficult for me to be honest with you—in the same way that it is difficult to be honest with my mother. I'm afraid of confrontations that are unpleasant. I've avoided them from the time I was a little girl.
>
> "In Europe, I had no desire to follow your plans and route and at the same time didn't have the courage to confront and tell you, and so the tension grew and grew between us, and finally I exploded, which is a definite pattern for me, or has been always. It was not just a question of following your plan, either, but also of being sort of psychically bound by your very strong personality. Somehow, I could never quite assert my real self around you. You were always in the lead.

"*From the time that we became pretty close friends in high school, the beginning of our junior year, you had a great deal of influence over me. You treated me wonderfully, and made me feel very special, like I was the best friend you could ever have. I thought you were such a fine, pretty, smart, religious, virtuous, and completely faultless person, and immediately made you my model and idol. I always felt inadequate, lazy, and guilty with you, but was also happy that you came along to be my image to try and emulate. I never had a huge opinion of myself, and now began to feel guilty about all the things (qualities) I thought I should possess, that you seemed to have and I didn't.*

"*I think I was afraid to ever defy you in any way, and had no conscious desire to. My mother used to say to me that you influenced me too much, and that I should do my own decision-making about things, and I used to argue with her endlessly about it. I see now that she was partially right about your strong influence. Your inability to take anything from anyone unless it was a formal birthday or Christmas gift used to infuriate me, but finally the sheer strength of those feelings of yours won out, and all I could do was feel guilty at every treat of yours, and feel miserable whenever you gave me an extravagant gift. And so much guilt and obligation all the time. I wanted to be free of it, but you could never accept anything from me, and you always won out on every score.*

"*When we were in Idaho with you, I saw so many of the same traits in Olivia, and it made me sick. These are not little meaningless personality traits, they are serious symptoms of illness, and now Olivia is being instilled with the same stuff your parents infected you with. You have tremendous guilt inside of you, and I really believe that you give mostly out of a huge, driving obligation feeling.*

"*When I look at you, I see that you don't love yourself at all. I see all your self-hate being transferred to Olivia and Lary. You insist that you are happy and that everything is all right. I'll never be convinced until you seek help, and it's getting later all the time for you. That is what I see. I'm sure that you must disagree just for your own mind's preservation.*

I am at a loss at what to say to you to make it all right and still tell you the truth as I see it. I think you are ignorant of a lot of the basic ways minds work. I was ignorant too, up until a few months ago. Your weeks with Bartelme were no doubt temporarily good for you, but believe me, it takes very slow, steady work to get at some of the really deep and basic things that help to destroy us inside.

"... I saw an environment of near perfection on the outside, and real pain inside. I have finally been able to look at you and say: "That is a sick person."

"The week before we arrived at your house was so full of anxiety for me. I was literally shaking like a leaf when we drove up to your house. It was like I was going to my own self-sacrifice, or something. I was fighting like crazy to hold on to my identity while I was there. I couldn't give anything to anyone. I think I was beginning to know a lot of the things inside me that our relationship had been based on, and the beginnings of rebellion against my own weaknesses. I was very unsure of why I was feeling so scared. It was the old threat of confrontation again. The little honesty I did manage to get out was all jumbled up with fear of hurting you, just fear of more pain for us all."

As if channeling Dr. Rosen, Nonny had declared I was hurting Olivia. Her sudden change of heart also triggered old physiological reactions to my mother's erratic demeanor. I had shared and entrusted more of my true self with Nonny than any other human being, and now she had suddenly re-written our shared history and relationship in an extremely negative and deeply disturbing way.

The day after Nonny's letter arrived, we were scheduled to drive to south to Boise for the accident trial. I was forced to pack up my despair over Nonny's retrospective rejection. Five days later, in a Boise courtroom, Attorney Peter J. Boyd won our lawsuit, but barely. In 1967 Idaho juries were suspicious of people from California who looked like us. The Idaho State Legislature, at the request of then Governor Don William Samuelson, had just voted for a "Kick a Beatnik in the Seat Week." With this backdrop, Charles Blanton, the lawyer for Keith Taylor and his father, did his best to make us out as drug-crazed transient hippies.

The jury responded by limiting our award to actual losses since pain and suffering are endemic in Idaho. After paying the remainder of the Caldwell hospital bill and Attorney Boyd's thirty-three percent contingency fee, Lary and I were left with about eight thousand dollars. We put most of it into the First Bank of Troy, a deposit which resulted in more favorable looks from Frank Brocke, the owner. Before that deposit, Brocke had always eyed us as if we were potential bank robbers.

While living on Nora Creek Road, we didn't know that in September 1963, three young robbers from California had held Frank Brocke, President of the Bank of Troy, his wife, and their fifteen-year-old son hostage in their home overnight. In the morning, the trio of robbers forced Brocke to open the bank vault before locking him inside, along with twenty bank customers and staff. Escaping with $55,000, the three robbers were soon arrested in California and convicted. The robbery and hostage taking was a shocking crime in Troy. Understandably Troy citizens were still viewing strangers from California with considerable suspicion.

When the settlement check arrived, I was still swamped by despair about Nonny's change of heart. At Lary's urging, I made a solo bus trip to Berkeley in November, a trip I described in a letter I wrote to my father on January 3, 1968.

> *"There is one other trip which I took right after we got the settlement. I didn't write you about because I was too upset. The week Lary wrote his letter to you, I took the bus to San Francisco for a four-day visit. What happened is that when Nonny and Jim were here in August, things were quite strained – for me they were strained because Lary's parents had just been here and because of our food expenses we were over-drawn at the bank before Nonny and Jim even arrived. We intended to ask them to help with the food, but they brought so many wonderful gifts that we couldn't bring ourselves to do it.*
>
> *I guess things looked worse to Nonny than they were, but after they left – about four weeks after, she wrote me a letter and said she thought I was a sick person and a lot of other things—she also said she didn't know what to do about our relationship. I won't elaborate, instead I'll just enclose her letter and let you see for yourselves. Please return it or bring it home with you.*

The things she said were so upsetting I became terribly depressed. I just couldn't seem to handle it. I finally wrote her telling her that I understood her concerns and that I hoped we would someday be able to work something out between us.

Then we received a letter from Jim saying he thought I didn't want to deal with the whole situation at all—neither Lary nor I could understand what Jim meant by this. At this point the settlement check arrived, so I took the bus to SF to see Dr. Bartelme and hopefully Nonny. I spent two hours a day talking to Dr. Bartelme about the letter and the things that were still bothering me. I was struggling to understand why I was so hurt and depressed.

I saw Nonny briefly the second night I was there. Nonny came over to Berkeley with Jim; while he went to his therapy appointment, we talked in a restaurant for two hours. It was not as awkward as you might imagine. In fact, I couldn't seem to get her to talk about what had happened between us at all. Nonny and Jim have bought one hundred sixty acres of land in Humboldt County in northern California near Garberville for seventeen thousand dollars, putting six thousand dollars down (the last of Nonny's inheritance from her grandmother) and she was very excited about this purchase, which was probably the reason I couldn't get through to her. Anyway, we parted with me feeling completely frustrated.

I called her later that evening and we went over the letter point by point and the things she said were even stronger, but more in relation to our past friendship. She kept saying things, but I just couldn't take them in or believe. About how she had always felt inferior to me and uncomfortable with me...that is, until the great Hodgkin's hoax and after that she had just hated me, except that she didn't know this until lately. She said she felt inferior to me artistically and this bothered her because she was supposed to be the artist and that she dreaded coming to each of the new places we fixed up because they made her feel bad... etc.

After I talked to her that night I felt very sad, but the next morning I couldn't remember the things she had said to me or else, if I did remember them, I found them hard to believe. I ended up talking to her two more times—once from Dr. Bartelme's office so that I could take down the things she said and we could talk about them. She said she had wanted to be free of me since our freshman year in college. That I had some kind of grasp on her. I was too strong, had a calculating, computer-like mind, a dominating personality and gave her a kind of love and devotion she could never return. She found me false because with her I was something other than that which the outside world saw. That is true, but I always thought she liked me for what I really was, not for awards or other things. Finally, Nonny said she didn't ever want to live within ten miles of me. Really she said she didn't want to have any kind of relationship with me.

Dr. Bartelme helped me, but the shock waves are still coming. I feel so empty sometimes and terribly nervous about making new friendships. Nonny has always meant so much to me, and I thought she liked me for what I was and this is why it hurt ...sort of the ultimate rejection according to Dr. Bartelme. I am doing better now. At least I have accepted the fact that our friendship is over, which is something I couldn't do at first. Even on the long bus ride home (forty eight hours) I found myself thinking of Nonny in the same old way."

The most important insight Dr. Bartelme gave me was something that I couldn't mention in my letter to my parents. Dr. Bartelme had listened to my last conversation with Nonny on the extension in order to hear her comments for himself.

After the call ended, Dr. Bartelme said, "Nonny has not only retroactively condemned your relationship, but also, by declaring that you are ill and infecting Olivia, she has unearthed your worst fear—that of becoming your mother. It may take you years to heal from her wounding words, but I feel certain after seeing you and working with you this week, that you will never develop schizophrenia."

-46-

The Gold Star Girl

In 1968, Lary and I settled into more comfortable and predictable routines of teaching, being taught, and maintaining the Brotman property. The latter was done with the devotion of my father. The house now had an attached cold storage room, an indoor toilet and shower, and the lower level interior was snugly insulated with cedar sawdust, covered by, what even Gilder agreed, were beautiful weathered barn boards.

At the end of our second summer on Nora Creek Road, one free of both accidents and company, Lary and I took stock of our Idaho sweat equity. We had explored the possibility of immigrating to Canada to start over again and had our passport photos taken in preparation for that possibility. But we concluded that we had learned enough to permanently settle in the midst of culturally hostile and geographically challenging territory.

I wrote Norah to ask if she had any interest in selling. She wrote back with an asking price of nine thousand dollars. Because we had just about that amount in the First Bank of Troy, I wondered whether Banker Brocke had called her with the figure. We sent Norah a deposit and prepared to finalize the purchase in October.

After our move to Idaho, Olivia and I heard nothing more from Dick, who had married Linda Pierce in Berkeley on June 16, 1967, my twenty-fifth birthday. Florence sent us the announcement, along with a handwritten note, which I read to Olivia. I was not surprised when she asked no questions. Olivia had only snapshot memories of Dick—the father she knew and trusted was Lary.

In contrast to Dick's silence, Florence wrote often, always remembering to send Olivia a small present for various holidays and for her birthday. Having been raised on a stew of secrets, I tried not to keep any from Olivia. I erred on the side of truth too often for her developmental needs, but that was part of the legacy she was inheriting.

I never said, "Your father loves you, even though he isn't writing."

When Olivia began asking to send some of her numerous drawings in the mail, the only recipients she designated were "Grandma Florence" and "Grandpa Frank."

On September 4, 1968, a large yellow school bus lumbered up Nora Creek Road and stopped on the road below Olivia's house. The door opened and Olivia struggled to reach the first step, waving goodbye from an empty a seat near the driver. Olivia's excitement about going to school and riding the bus had been palpable for weeks. Until 1968, Idaho did not have kindergarten programs in its public schools. But thanks to the War on Poverty and the Elementary and Secondary Education Act (ESEA), conservative states like Idaho had begun accepting federal funds for early education efforts.

The teacher hired for the first kindergarten class at Troy Elementary School was Miss Hermine Benson. Her prior teaching experience was limited to two decades of work at the Lutheran Sunday School, where she had honed classroom management skills heavily dependent on acts of banishment, shame, and gold stars.

It was several weeks before Olivia recognized the futility of her efforts to earn a gold star or some level of acceptance and approval from her first teacher. In a class of eighteen kindergartners, Olivia was the only girl who had not earned a single star, a deficit recorded on a large chart posted on the wall. Two boys were also without a single star.

Like many of the long-time residents of Troy, Miss Benson did not look kindly upon strangers. Olivia's demeanor was different. Already a confident reader, her vocabulary was advanced. My parents had sent Olivia two coats and a half dozen dresses from Europe which, although beautiful, marked her as an outsider. In Troy, like all small towns and isolated communities, any stranger was grist for the gossip mill. Had we began attending a local church, attitudes might have softened. But after two years, the only friend we had east of Moscow's city limits was Glenn Gilder.

During the first weeks of school, whenever Lary or I asked, Olivia expressed mild ambivalence about Miss Benson, but she never came home in tears or announced she did not want to go to school. Instead, Olivia woke early, dressed herself, ate quickly and waited at the roadside before the sounds of the approaching bus could be heard echoing through the forest.

During the fourth week of school, Olivia decided she wanted to eliminate her Gold Star deficit. Students went to the Troy Market, a half block down the hill, for gum or penny candy during lunch and after school. School security was non-existent in 1968. Troy Market served multiple needs; among the school supplies were match-box-style containers of gold stars. Instead of sitting down to eat her sack lunch with her classmates, Olivia went to the Troy Market and purchased two boxes of gold stars. The classroom was still empty when Olivia returned, so she quickly implemented her premeditated plan to achieve a measure of justice for herself and her two starless male classmates.

Olivia in platform rocker after first day of kindergarten in Troy, September 1968

As Olivia later described the scene to Lary and me, Miss Benson did not immediately notice the improved star-status of Olivia and the two boys, but her classmates did. Some began whispering and giggling. Olivia remained silent, carefully observing her teacher. When Miss Benson recognized that unauthorized changes had been made, she roughly jerked the chart from the wall. Holding the chart close to her chest, she briskly walked out of the classroom, leaving a trail of gold stars floating behind her.

Long minutes passed before Miss Benson returned, flushed and empty handed, to announce, "There will be no afternoon recess until those who defaced the chart come forward to accept responsibility for their deviant act."

Miss Benson assumed an act of this nature could only have been committed by a group, not a single individual. Olivia and her classmates remained silent through the afternoon recess period, listening to the sounds of exuberant physical release on the playground outside the windows of their tense classroom.

Olivia kept her secret until she returned home. Then after asking us what "deviant" meant, Olivia described the drama of her day. Lary and I told her we were proud of her creative courage, but she needed to tell her teacher what she had done so that her classmates wouldn't continue to be denied recess.

Olivia had an additional worry, she said, "I'm afraid Miss Benson won't replace all the gold stars that fell off the chart when she took it out of the room. I don't want others to lose stars because of what I did."

The next morning I left for WSU before the school bus arrived in order to arrange a conference with Miss Benson and the principal. When I returned in the afternoon, the first thing I saw was Olivia sitting on an adult chair outside the principal's office. She jumped down and ran to embrace me. The grocery store clerk who had rung up Olivia's lunchtime purchase was the brother of Miss Benson. Olivia had been taken to the "In Trouble" chair the minute she entered Miss Benson's classroom. The conference did not last long.

The principal sheepishly explained that Miss Benson was adamant she could not continue to teach with Olivia in her classroom. However, he reported that Miss Benson felt Olivia was "mature enough" to move up to the first grade classroom. Otherwise, the only option he could suggest was that we consider "trying" to enroll Olivia in a Moscow kindergarten.

"In general, your family does not seem to be a good fit for our community!" he said without looking directly at us.

Silent tears began spilling down Olivia's cheeks.

My mouth turned dry and cottony as I asked, "Are you saying that Olivia is being expelled from Miss Benson's class?"

"No, I am saying that there is no longer a place for her in Miss Benson's class!" he responded.

"I find that distinction fictional," I said. "The removal of a five-year-old child from a federally funded public education program in order to meet the demands of an unlicensed and clearly inept teacher is shameful. I am not even certain that you can do so under the terms of your federal grant."

After telling the principal we would consider our options over the weekend, I walked Olivia down the hill to the Pontiac. She was understandably distraught as we drove home, but by Saturday morning, she had reclaimed her normal equanimity, saying she was willing to change classrooms, as long as she could see her former classmates at lunch, recess, and on the bus. What I didn't immediately realize was that Olivia had a new status—she was already a kind of heroine among her fellow students. Miss Benson had not earned a single gold star on anyone's chart.

Olivia made a successful transition into the first grade class. A few weeks later, we all overslept after having come home late from a celebration of Lary's birthday with the Sullivan family in Moscow. While the bus idled loudly, Olivia quickly dressed and I made two peanut butter sandwiches—one for lunch and one to eat on the ride to school. Before running out the door, Olivia grabbed her coat from the floor where it had fallen from its hook. From the kitchen sink window, I watched her pulling her coat on as she ran down the slope to the bus and climbed aboard.

By the time the bus reached the school, Olivia was in real distress. During the night one of Grey Cat's kittens had made a small bowel movement on the lining of Olivia's coat sleeve. The first grade teacher was not nice about Olivia's situation. After roughly removing her dress and wetting a brown paper towel for Olivia to use on her shoulder, she gave Olivia an abandoned boy's t-shirt to wear on top of her full slip, before sealing the dress and coat in a garbage bag.

When Olivia stepped off the bus and climbed the slope to the house, she opened the door and announced she would never go back to Troy Elementary School again.

As luck would have it, Olivia didn't have to face the problem of re-entry. The next day my parents returned from Europe. They had sailed home on the same boat as their Beetle and driven across the country.

After spending a few nights with us, they decided to rent a furnished apartment in Moscow on a month-to-month basis. Olivia loved going to stay with Grandpa Frank whenever I was at work or school, which is how Olivia became home-schooled: Lary handled art and music; I took on science, cooking, and French; and my father handled Olivia's curriculum for math and reading. Even my mother helped by making a princess costume for Halloween so Olivia could have her first experience with the power of trick or treating on the streets of Moscow.

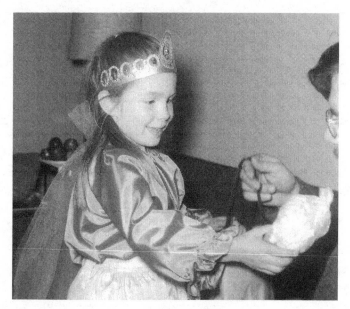

Olivia showing Lary her Halloween Treats, Moscow, Idaho, October 31, 1968

It was during this period that my father saw a television program on testing for lazy eye before children started school. The test was simple. Taking one of Olivia's books, he covered her left eye and asked her to read him the letters. When my father covered her right eye, Olivia panicked and burst into tears because she couldn't see. That was the moment we discovered the permanent injury she had suffered in the truck accident. The force of her landing in the muddy clay basis had severed her left optic nerve. Filled with shame for not having discovered the injury sooner, I wept at her loss. Olivia remains blind in her left eye.*

We left the house on Nora Creek Road forever just before Thanksgiving. Our efforts to make a life on the edge of a Northern forest disintegrated during the title search for the Brotman property. The search revealed that Norah had signed a quit claim deed in 1960 when she divorced her husband, Jordan Brotman. The property belonged to him, not Norah—a legal detail Norah had forgotten when she offered to sell. We contacted Jordan, then teaching at

---

*Four months later Olivia became the patient of Dr. Firmon Hardenburgh, a Harvard trained eye specialist.

Sacramento State College. Still full of post-divorce anger, he was unwilling to sell at any price. Worse yet, he asked us to move out before Christmas.

A few weeks before the potential sale came to an abrupt halt, I had received an offer to join the War on Poverty, based on a referral made by former University of California President Clark Kerr. My assignment would be with Project Head Start in Boulder County, Colorado. We were about to be homeless, so I accepted the offer. I resigned my WSU position; we withdrew from our UI courses; and replaced the Chevrolet with a used box delivery truck in which to move our now larger collection of household goods and animals.

The night before we planned to leave for Colorado, my father picked us up at the house to spend the night at their Moscow apartment. Lary and I were filthy after a day of packing and cleaning out the house. We needed a place to shower and sleep, having shut down the well pump and drained all the pipes for winter. Just before climbing into my father's Pontiac, Lary decided to prop open the door to the cold storage room to reduce the remaining scent of our summer onions, potatoes, carrots and beets. We had given our stores of hard-raised produce to Gilder earlier in the day.

Pablo spent the night with us on the living room floor of my parents' Moscow apartment, but we left Grey Cat and Tiger Cat at the house—locking them inside with food, water, and a litter box because clearly our packing had disturbed them. We planned to load the two felines into the box truck the next morning when we returned with new combination padlocks for the house, shed, and well, and to pick up the fully-packed truck.

At dawn, Lary and I dressed in silence and quietly left with Pablo. Olivia was still in dream land. Driving toward the house on Nora Creek Road for the last time, a wave of deep sadness washed over me.

As we came into the vale, we caught a glimpse of a huge mother bear with twin cubs moving up the hill toward the pond and the forest. Inside the cold storage shed some of the cedar sawdust insulation had been spilled from the walls. The condition of our front door was remarkably similar to its appearance after our first bear visitor.

Stepping inside the house, I gasped. Several pieces of Tiger Cat's bloodied fur coat and one of his paws were on the floor near the animal's food bowls. I could hardly breathe thinking of trying to help Olivia through another pair of feline losses. Weak-kneed and teary, I started up the stairs when I heard a faint cry. At the top, I heard the same sound again. Then I saw Grey Cat peek her head out from the secret nest Lary had made for her kittens between the rafters in Olivia's room. Grey Cat was safe!

The House on Nora Creek Road, Fall 1968

At the news of Tiger Cat's death, Olivia sobbed and held Grey Cat close. Her beloved cat purred loudly, licking the tears flowing down Olivia's cheeks. My father generously offered to make the necessary bear repairs so that we could leave as scheduled.

Assuring us, he said, "I'll enjoy being out there and doing that kind of work!"

We had planned to wind our way down to Boulder, Colorado via Yellowstone. As soon as Olivia understood that Yellowstone was a place where bears lived, she sobbed until we promised not to go there. Instead, we took an alternate route through Logan, Brigham City, and Ogden. There we connected with I-80, crossing the southern half of Wyoming in the dark, seeing only antelope, jackrabbits, and one lonely coyote.

I had no desire to stop to see anyone or anything in Utah. Instead, I was anxious to reach Colorado. For the first time in my life, I would be living on the eastern side of the Rocky Mountains, but once again, Thanksgiving would be the demarcation for another phase of my life.

-47-

Coming Together, Coming Apart

Boulder looked magical nestled up against a forested mountain with its Flatirons pointing toward a bright blue heaven. The combination of civilization and wildness thrilled me then and still does.

Flatirons on Mountains, Boulder, Colorado

On the last Monday in November 1968, a week before I was due to report for duty in the War on Poverty, we found a one-bedroom recently restored carriage house for rent behind a Victorian mansion on Pine Street. The monthly rent of three hundred dollars was well above our imaginary budget, but the carriage house was clean, charming, and required no lease.

Two days later, we celebrated Thanksgiving at the Hotel Boulderado, a historic edifice in downtown Boulder just a short block from our new nest. As we were seated in the beautiful dining room, I felt we were home at last.

The next day Lary started looking for a used car for me to drive to work, while I searched for a school for Olivia. My faith in public education had been shaken. I wanted something safe, preferably a full-day program since I felt certain my War on Poverty work hours would be long.

Upland School,[35] an alternative school, had opened a year earlier in the Quaker Friends' Meeting House on Upland Avenue. The teaching style was low-key and relaxed. While the curriculum was thin, there were no gold star incentives. It seemed the best place for Olivia to regain her confidence in school. I bit into my share of our settlement grubstake to pay the modest tuition for Boulder's first "alternative school."

While I shopped schools, Lary located a 1956 Beetle for one hundred fifty dollars, which a mechanic at Modern Specialists promised he could keep running for another year. This early model was the essence of simplicity—the only gauge on the dashboard was a speedometer. Whenever the Beetle ran out of gas, I simply moved my foot to a lever on the right side of the gas pedal, pushed it downward, and two more gallons emptied into the tank. I drove that Beetle for five years, often commuting one hundred miles daily between five Head Start Centers in Boulder, Lafayette and Louisville. The bottom of the 1956 Beetle chassis was worn so thin that the roadbed was visible through the floor boards.

On my first day of service in the War on Poverty, my deployment orders came from the Office of Child Development (OCD) in the Department of Health, Education and Welfare, then the home of Project Head Start: "Create a parent-controlled non-profit corporation to act as the grantee." OCD was dissatisfied with the manner in which the Boulder Valley School District (BVSD) had been managing the county-wide program, and particularly with the level of parent involvement and the nutritional components. The BVSD grant had been cancelled mid-year, causing a local political storm. Temporary responsibility for Project Head Start had been transferred to the local board of directors for the Office of Economic Opportunity (OEO) until a new parent-controlled non-profit organization could be established.

The transfer of responsibility for the program was made over Christmas break. I had less than thirty days to arrange for the transportation, food, and maintenance services previously provided by BVSD to the five Head Start centers: three in Boulder, one in Louisville, and one in Lafayette. While the entire teaching staff agreed to continue in their previous roles, they were understandably uneasy. Several were openly suspicious of me. For the first two weeks of January, we were short one bus driver so I had no choice but to temporarily fill the role. I drove the YMCA's sixty-passenger bus, a vehicle that necessitated I obtain a chauffeur's license, but which suppressed my car hormones.

My parents became homeless migrant retirees. At the end of their Moscow pre-paid rent period, they left Idaho, my father driving the Pontiac and my mother following in their new Beetle—the first of many property divisions to follow. They stopped in Salt Lake City to visit family members during the Christmas holidays, before trailing each other to Phoenix. They rented another furnished apartment in Phoenix, the home of my mother's brother Jack Campbell Beck. My father took an art class and discovered that he had previously unknown artistic talent. My mother did well for a few months, but in April she stopped taking her medications and the voices of fear and paranoia returned at full volume.

Their European travels had been driven by my mother's paranoia. She believed a French woman was following them through Holland, France, and Italy. Giant cockroaches in Greece were really FBI "listening devices." Spotting a German man, who seemed to be following them on the Swiss border, my mother became convinced he was a spy. Plus there was a panoply of real surveillance in both Russia and East Berlin.

Their plan had been to rent small apartments for month-long stays in order to explore in depth. My mother's fears were so controlling, that they never stayed more than a few nights in any one location. She also insisted they travel to Israel in the middle of their trip. This destination reactivated many of her old ghosts. Once again, she became preoccupied with the complex relationships between Jews and Mormons, as well as between the State of Israel, Palestinians, and the United States.

The director of Boulder County's OEO coincidentally resigned on my first day of service as director of Project Head Start. He offered me a personal economic opportunity—the assumption of the lease on his rented cottage. While he did not offer an inside preview, Lary and I drove by what he said was a six-hundred-square-foot cottage located at 751 Marine Street in West Boulder. As recent immigrants from Northern Idaho, the cottage appeared more than adequate. The rent was only one hundred dollars a month, the first within our hoped-for-budget. Built in 1909, the cottage was within walking distance of downtown and the University of Colorado. Most importantly it had a much-needed fenced yard for Olivia, Grey Cat, and Pablo. The next day the director took us to meet his landlady and we assumed his lease.

On January 3, 1969, the director left the key to the cottage for me at the OEO office. After work, Lary and I opened the front door, while Olivia romped with Pablo in the yard. Stepping across the threshold, I felt as if I had fallen into purgatory.

The director had gone out of town over the Christmas holidays. Unable to find a dog kennel willing to board his two adult Irish Setters and six puppies together, he had instead left them alone in the cottage. His provisions for water and food were made by leaving the bathtub tap and the toilet running, along with a large open bag of dry kibble. The house was uninhabitable, the floors covered with excrement and urine.

The landlady gave us free rent for three months in exchange for Lary's sweat equity and skilled labor. Fortunately we were able to remain in the Pine Street carriage house until the Marine Street cottage was habitable, an effort requiring all of Lary's time and talents. The linoleum had to be removed, wallpaper steamed off, and the baseboards and hardwood floors replaced. As usual, Lary worked extremely hard in his self-supervised, unsalaried position. He winterized a small porch off the kitchen to serve as Olivia's bedroom, doubling its floor space by building a loft and creating storage and play space beneath her bed. Slowly, the cottage began to seem as if it would be a safe place for all five of us.

The day after we moved into the Marine Street cottage, we took the box truck into downtown Denver in search of a couch. I wanted the couch drought to be over forever; Lary and Olivia agreed. We had heard that Denver's secondhand stores had a better selection than those in Boulder, which were overused by Boulder's twenty thousand university students. In in a Volunteers of America store on East Larimer Street, we found a matching couch and reading chair in acceptable condition, priced at ten and five dollars respectively. The couch appeared to be a twin of the one my parents bought at Dinwoody's in 1941— rust-colored floral damask with plump cushions and soft fat arms. I felt overwhelmed by the coincidence.

I saw Olivia watching me in the mirror with wide eyes and remembered my mother's magic Christmas dress shopping in Ogden. Except for one seven dollar summer shift, a well-worn raccoon coat from a Caldwell thrift shop, and a pair of winter boots purchased after the Boise trial, these were the first street clothes I had acquired since Olivia's birth. The total price of the couch, chair, and restorative wardrobe was twenty-eight dollars. Unbelievable!

The Marine Street Cottage, Boulder, May 1969

Several months before we decided to leave Idaho, I had been examined by a gynecologist in Moscow. Not only had I stopped menstruating, but my hair was repeatedly clogging the shower drain. The male Moscow gynecologist looked normal enough, but he was angrily adamant that I immediately stop taking the first generation of birth control pills, which I had been using since Olivia was weaned in Berkeley at seventeen months.

In addition to discovering that my thyroid function was dangerously low, the Moscow gynecologist had another warning, "Do not take another birth control pill the rest of your life. The ones you have been taking have shrunk your uterus to infantile dimensions! I am very sorry to say this, but you will never again be able to conceive a child."

Hesitant to accept his prognosis, I asked, "Shouldn't I use some other kind of protection, like one of the new uterine devices?"

The gynecologist responded by pulling open the bottom file drawer of his desk to show me what he thought of what we now call IUDs. It was overflowing with ones he proudly claimed he had removed.

"These are more dangerous than birth control pills—don't ever allow one to be put in your uterus!" he exclaimed.

I left his office with a prescription for replacement thyroid hormone and a strange sense of sadness, mixed with relief. I had never considered having another child, but any line drawn in the sand of one's future is disturbing, sometimes provocative.

Three months after our move to Boulder, I had the second gestational shock of my life. The theory I later developed to explain my own restored fertility was that the Moscow gynecologist was a Northern Idaho survivalist in disguise, many of whom believed women should be kept barefoot and pregnant. He had used his medical scare tactics in furtherance of his regressive vision. Either that or else he was just an ordinary quack.

Despite repeated corrections, my Boulder gynecologist always addressed me in the diminutive.

"Joanie," he said, "I think this might be the wrong time in your life to have another baby. Your uterus is too small. I believe you are likely to have a miscarriage or serious complications during your second trimester."

There had been some progress in women's reproductive freedom since mistletoe failed Nonny. In 1969, in Colorado, it was possible to obtain a therapeutic abortion in a hospital provided three physicians or psychiatrists vouched for its medical necessity. I made the decision to abort because my doctor felt it was best. After minimal discussion, Lary concurred. I called my Boulder gynecologist to tell him to make arrangements for the procedure, which, in those days of luxurious healing hospitalizations, involved an overnight stay and anesthesia.

Coincidentally, the same day we decided to abort the pregnancy, Lary received a letter from Jim Ekedal forwarded from Troy. Jim wrote that he and Nonny were about to leave San Francisco in search of a new place to live in either Iowa or the Ozarks. Lary wrote to Jim, giving him our new address and telephone number.

Jim replied with a request: "Could the two of you meet us in Taos for a reunion?"

I went to Taos reluctantly, in a highly defended state. The four of us and Olivia spent a day talking while exploring. Because Nonny was in a much different emotional state, it seemed as if we were able to find our collective way back to our Berkeley-way of relating. For me, however, something was fundamentally different about my relationship with Nonny. Being near her made me feel as if I was walking on thin ice—every word I spoke, every gesture could plunge me into her icy waters.

Coincidentally, Nonny was pregnant for a second time, this time with Jim's baby, not Bill's, while I was pregnant with Lary's, not Dick's. In the face of

Nonny's positive and hopeful expectations, I made no mention of my pregnancy or planned abortion, a secret that increased my wariness.

After our Taos reunion, I caught the perpetual Head Start cold; it escalated into bronchitis. Reaching the limits of the law, I had to proceed with the abortion, developing an intractable walking pneumonia from the anesthesia. I was still hacking away when Nonny and Jim made another appearance on our life stage.

They knocked on our Boulder door in late March in search of temporary shelter. Their eastward exploration had been driven by Nonny's fear of earthquakes. She had read the writings of Edgar Cayce,[36] swallowing whole Cayce's prediction that at any moment California was going to slide into the Pacific. Not finding either Iowa or the Ozarks to their liking, they decided to try Colorado.

Early each workday morning, as I gingerly stepped over the slumbering bodies of Nonny and Jim, trying to reach our only bathroom, I wanted to take Olivia and move into a nearby motel. The chaos and crowding in six hundred square feet of space was not my idea of *optimal distance*. Further, Nonny and Jim had transformed into an archetypal Age of Aquarius couple preparing for natural childbirth and were fully focused on practicing their Lamaze breathing exercises. Every day they studied the horoscopes published in the *Boulder Daily Camera* and the *Denver Post*. Lary, still extremely susceptible to the nearest cultural virus, fell under their astrological spell; he too began relying on his horoscope for daily guidance. A few months before the arrival of Nonny and Jim, Lary had begun to use his architectural inclinations and carpentry skills for part-time remodeling work. Jim proposed working with Lary and they christened their astrologically-controlled partnership "Zodiac Construction."

Lary finally helped the star-guided couple move into a rental cabin near the mountain town of Nederland, west of Boulder. After a month in the cabin, the soon-to-be parents moved to a farmhouse near Haystack Mountain northeast of Boulder. I was both too busy and too short of breath to be of use for either move.

I ended my service as a Vietnam War Bride early on a June evening. I was watching Olivia practice cartwheels on the front lawn, trying to muster energy to prepare dinner, after a day of packing up the Lafayette Head Start center equipment for summer storage. Lary came home from helping Jim paint the farmhouse and sat down next to me on the front steps.

Taking my hand in his, he said, "JC, I think we should have a baby together!"

Still in post-abortion progesterone tumult and seriously short of serotonin, depressed, and sleepless from constant coughing, I couldn't believe I had heard Lary correctly.

"What did you just say?" I asked. "Are you trying to make a joke?"

When Lary confirmed that he thought we should have a baby together, something snapped inside me. For months, I had felt as if I was the only responsible adult among the four of us. Not only had I been paying the rent and buying the food, I had a low tolerance for the visions of Edgar Cayce. More importantly, watching the three of them being guided each day by horoscopes was too close to my mother's form of message madness. After several minutes of silence, I stood and asked Lary to come inside, out of the earshot of Olivia.

My voice trembling, I said, "Lary, I have reached the end of my heroic need to save you from either death by war or death by your own susceptibility to the Doom's Day and New Age thought-processes of others. I no longer want to be your wife for any reason, and the best one yet is the timing of your latest proposal to have a baby while I am still trying to recover from an abortion we agreed upon!"

In response to Lary's silence, I tried to find calmer water and said, "Olivia needs summer sandals. While I take her shopping, I would like you to gather your essentials and to not be here when we return. When I am less angry, I hope that we can talk to Olivia together."

Lary appeared to be in a state of shock and said nothing.

I opened the screen door to tell Olivia to wash up for shopping and dinner at Round the Corner.

While Olivia was in the bathroom, Lary said, "I don't understand you, but I will leave if that is what you want."

"That is what I want," I replied, while wanting to pound his body with the full force of my fists.

Later that night, Olivia and I returned to the empty cottage. I told her that Lary had gone to stay with Nonny and Jim.

"When will he be back?" she asked.

"I don't know the answer to that question, but I am certain you will see him again very soon."

That night both Olivia and Grey Cat crawled into bed with me.

As we lay in the dark, Olivia asked, "Do you think Lary will remember to feed Pablo?"

She knew there had been a real break.

When Lary left, Pablo went with him. It was right. Lary was Pablo's alpha human; I was working full-time, and Olivia was at the YMCA day camp. Olivia missed Lary more than Pablo; I missed Pablo more than Lary. Olivia was old enough to be able to express her sense of loss. Sad and subdued, she asked the same questions about Lary over and over.

As my hormones returned to normal, Olivia's visible psychological wounds caused me to reexamine my decision. I had approached both of my marriages as if they were temporary dead-end small business propositions. The purpose of the first had been to bring Olivia safely into the world. The purpose of the second had been to keep Lary from being killed by the Vietcong or, in retrospect, perhaps being brought up on a court-martial by his commanding officer. Neither marriage had been influenced by the marketing of enchantment. Sexual relations were non-existent with Dick and a secondary benefit with Lary. What I didn't want to lose was what Lary had provided to Olivia as a father.

After several weeks of separation, Lary stopped by the cottage to discuss our situation. Grey Cat greeted Pablo with feline joy. Standing on her hind legs, she began licking his face. Pablo lowered himself to the floor and Grey Cat snuggled into his warm belly.

Not only was I responsible for hurting Olivia, I had also split apart two loving animals. I felt terrible. Looking at Lary sitting on the twin couch, our two beloved animals reunited at his feet, I felt a wave of nostalgia for all we had shared. More importantly, I had searched desperately, but unsuccessfully, for something with which to treat Olivia's melancholy, knowing I had done deliberate damage to her psyche. Repairing my marriage to Lary was my only option. I brought Lary a glass of orange juice, as if he were a guest in his own home, and tried to keep my voice gentle.

"Lary, I was a hormonal mess when you proposed having a child together," I said. "Please accept my apology for my angry reaction. Would you be willing to give our marital arrangement another try?"

Lary quickly interrupted, "J.C., one of the reasons I stopped by tonight was to tell you I am living with another woman!"

Something about Lary Carpenter was irresistible to most women.

I couldn't sleep that night; neither hot fudge, nor music smothered the shame. In a moment of uncontrolled anger, I had failed my beloved Olivia by impulsively cutting away the only father she had ever known and permanently breaking apart a powerful cross species relationship.

My lesson in love: Human hormones are the most powerful ingredients in the stew of cells, proteins, and molecules that comprise the human species. Sources of enchantment, joy, rage, pleasure, depression, and sorrow, the influence of hormones seems to determine who lives and who dies, who stays and who goes. Hormones, not opiates or amphetamines, should be controlled substances, carefully doled out under specialized medical supervision.

The next day, I obtained a cashier's check in Lary's name for more than half of what was left of the hay truck accident award, seeking unobtainable solace by giving him more than his fair share. Check in hand, I met Lary at the cottage while Olivia was still at day camp. I did not want Olivia to witness Lary's departure, deluding myself that it would protect her from the finality of loss.

Pablo and Grey Cat watched us both carefully as we loaded Lary's clothes, tools, and household goods into the box truck. Lary gave me a neutral hug before he drove away, heading west on Marine Street. Pablo put his head out the open passenger window and looked back at me as I stood on the sidewalk holding Grey Cat in my arms. Pablo's head, tilted to signal he was listening and struggling to understand the behavior of his human companions. Pablo's beautiful big brown eyes dissolved the Ben-Evans-style equanimity I had been struggling to manifest. Still holding Grey Cat, I sat down on the front steps and began sobbing. She moved toward my neck and began licking my tears.

The remainder of 1969 was painful, filled with a sense of emptiness, loneliness, and failure. I tried hard to fill the hole in Olivia's heart, but it kept springing leaks. I doubt Lary Carpenter ever looked back.

Zodiac Construction had an astrologically-determined slow start, and soon went into sleep mode. Lary and Jim reluctantly found work in a furniture factory, shortly before Nonny gave birth to a baby daughter at the end of September. Lary's new woman, Stephanie Lusak, began spending time with Nonny at the Haystack rental house while the men worked. I met Stephanie when she came with Lary to pick up Olivia for a weekend visit. I felt no malice toward her, but her presence naturally increased the distance between Nonny and me.

In January 1970, Nonny and Jim, along with Lary and Stephanie, left Boulder, joining the trendy "back-to-the-land" movement, having decided to live communally on the Humboldt County acreage Nonny and Jim had purchased in 1967. Nonny's Cayse-inspired earthquake worries had begun to recede, tracking a similar decline in her pregnancy hormones.

Lary and Stephanie drove the box truck, first to Bakersfield, and then to Humboldt. Nonny and Jim followed in their own vehicle. Nonny and her baby daughter stayed with her mother in Bakersfield until the others had established enough of a structural foothold on what was soon known as "The Land" for Nonny to feel assured her baby daughter would be safe and warm.

In August, Olivia and I flew to California. We spent three days in the Bay area before I drove Olivia to The Land for a two-week stay. When I picked her up she was glowing and full of adventure stories, having slept in a hammock hung from a tree already inhabited by tree ants.

Olivia continued to pay annual summer visits to The Land until she was fourteen years old, where she was welcomed and lovingly cared for by both couples.

Olivia's first visit to The Land in Humboldt County, August 1970. Left to right: Lary, Jim, Olivia, Nonny, and Stephanie Lusak (between curtain), and three other residents of the area.

Four years after our separation, I filed for divorce from Lary Carpenter. Although he was served papers in Humboldt County, he failed to respond. The final divorce decree was issued by the Boulder County District Court in February 1974 and my maiden name was restored.

Pablo continued to be devoted to Lary and Lary to Pablo. On a hot summer day in 1976, Pablo sought shade beneath Lary's pickup, while Lary was helping Jim with the construction of his house. A few hours later, when Lary climbed back into his truck, he whistled for Pablo, not realizing Pablo was sleeping beneath the truck. As the truck rolled forward, the rear wheels crushed Pablo to death.

Stephanie remembered this about Pablo, "Did you know that Lary skinned Pablo, in order to keep the hide as a remembrance of him? We kept his pelt in our house and used it like a couch pad or blanket to lay on. I was kind of horrified actually, but grew to appreciate it for what it meant to Lary. I loved Pablo too, but I never could have skinned him. That was Lary."37

Nonny and author, San Francisco, California, 1979

My friendship with Nonny continued after her devastating written and oral critique of my character, although for me the distance between us was never the same. It was greater, which I believe she had been seeking since our first year of friendship.

Nonny's kindness to Olivia during her annual summer visits to the Land helped repair and establish new boundaries in our friendship. We sometimes spoke on the telephone when Nonny had access to one and she made several visits to Boulder. Once we had a mini reunion in San Francisco when I was there on a consulting assignment.

In 1976, Nonny went through a period of intense attraction to Lary. She told me about their brief affair and her longer struggle to suppress her desire for him in letters and telephone calls, but I found myself not wanting to know the details.

Sadly, Nonny's marriage to Jim ended in divorce in 1979, not because of Lary, but because Jim, (then known locally as "Deer Hawk"), fell in love with a younger woman (known as "Sun Song").

Just as in romantic love, the distance we keep between ourselves and a friend changes, sometimes suddenly. My own experience is that the loss of a friendship can be more painful than the end of a marriage. Yet, there are no healing rituals or court proceedings to certify the change.

-48-

The High Price of Help

It is hard to help another person. It requires an intuitive sense of how far to move into someone else's space, as well as when and how to withdraw. My desperate efforts to help my mother had boomeranged, just as the help I received from others deserved mixed reviews. I repeatedly learned that even well-intentioned help is not always helpful. With friends I often found myself crossing boundaries without adequate permission, but in work-settings my instinctive helpfulness was almost always valued.

As a single parent with a six-year-old daughter, I was in charge of a complex and highly demanding social change effort. Every day I felt at real risk of failure. When I assumed the directorship of the Head Start program, all five of the Head Start centers were located in Protestant churches where ninety-eight percent of the congregants were white, even though the majority of the students were Hispanic Catholics or black Baptists.

I moved the three Boulder centers into houses located in minority and low-income neighborhoods. The two centers in Lafayette and Louisville were moved into buildings formerly occupied by restaurants. This gave each center a kitchen so the nutritional aspects of the curriculum could be implemented, instead of busing in largely wasted food from Boulder Valley school cafeterias. Consistent with my cultural conditioning, I moved my office into a Quonset hut located on a parcel of land designated to be a park on Canyon Boulevard near the Goss-Grove neighborhood, then one of Boulder's last predominately low income neighborhoods.

Olivia and I made it through the first year without Lary because Olivia met her first best friend, Zoe Young. Olivia and Zoe were both new students at Upland and strangers in Boulder. Coincidentally, Zoe's mother, Arlene Garber Young, became one of the Head Start teachers in the Boulder Valley program. Her husband Steven was a member of the CU neuroscience faculty. Between us, Arlene and I managed to get Olivia and Zoe to Upland School each day and to pick them up on time, or close to it.

On most days the two girls played together after school, usually at Arlene's, whose workday was shorter than mine. Whenever possible, I would arrange my schedule so that all I had to do later in the day was write or make calls. On those days, I too could be one of those invaluable after-school moms making a snack in the kitchen, producing the perfect dress-up accessories at just the right moment. While such moments were enriching, my sense of competence was much greater in my professional work role than as a mother because the shadow of my mother was so dark and so long.

The other mother in our all-female household was Grey Cat. Her first Colorado litter of three sweet kittens included a red-haired female, a female tabby, and a black male with a half-white face, chest, and paws. The male had a rabbit-like stub of a tail, just like his Idaho uncle, Tiger Cat. Olivia named them: Red, Tiger Two, and Manx. Olivia chose the latter name after we learned there was a breed of cats from the Isle of Man, called Manx cats with similar stubs. As the runt of Grey Cat's litter, Manx was very short, not only on tail, but also on trust. The diffidence of Manx appealed to Zoe, who begged Arlene to let her bring him home.

Arlene hesitated, "Okay, but not until after we come back from our summer vacation."

Olivia and I agreed to board Manx in the interim.

The next day my parents arrived for an uninvited and unannounced visit, having driven to Boulder from Phoenix in the same car, the Pontiac. Their arrival coincided with the busiest work week of my life. I was crazed and so was my mother. Of her own volition, she checked herself into Boulder Community Hospital their first the night in Colorado. She had picked up Mexican amoebas from something in the Nogales water, confirming her persistent fear of contaminants.

When I entered her double hospital room, my mother sat up and began raving, "You put the poison snakes in me again! God will punish you!" causing the mouth of the patient in the bed next to hers to fall open.

I felt sick at heart that my mother was once again in such close proximity, and did not allow Olivia to go to the hospital. The next day my mother's roommate was moved down the hall.

Two days later my father went to the hospital after dinner while I stayed home trying to finish an overdue OEO report. The internist treating my mother's amoebas had already called for a psychiatric consult and she was back on Thorazine, suddenly quiet and submissive.

After the hospital visit, my father picked up Olivia at Zoe's house where she been invited to have dinner. Parking in front of the cottage, my father didn't see Red sitting in the gutter. When he realized Red had been crushed to death, Frank cried more than Olivia, blaming the Pontiac.

"You can never see around the front wheel base. It's too damn big!" he said over and over.

We buried sweet little Red under the cherry tree in the back yard. The following week we found a home for Stripe Two, while Grey Cat and Manx settled into a mother-son dyad. When the Young family returned from their summer travels, Arlene decided she really didn't want a cat. Manx remained with us by default, the only male in an all-female household.

Thanks again to that unbelievable federal health insurance, the fifteen days my mother spent in Boulder Community Hospital gave my father ample time to find ways to help. He made myriad repairs in our cottage, drove Olivia to and from the YMCA summer day camp, went to the laundromat with our clothes, vacuumed, and groomed the garden.

My father made short daily visits to see his wife in the hospital, but those visits were merely transactional. Despite the Thorazine, my mother wanted all of her possessions brought to her hospital room to prevent the FBI from sneaking into their motel room or breaking into the Pontiac. She was certain agents wanted to plant listening devices in her clothes, books, and papers.

Within a few days of their arrival, my father had made up his mind that he and my mother would move to Boulder, his expressed reasoning: "We need to be here because you need my help and Olivia especially needs me!"

Olivia was close to her "Grandpa Frank" from the beginning, as if they had spent her whole life together, even though their total time together in six years had been less than six weeks. My father took his usual posture with respect to women with children who lacked husbands.

"You can't try to do everything yourself," he argued; "It is too much to manage on your own!"

His philosophy was reality based. From age seven, he had grown up in a family headed by an impoverished single mother with four children, who had moved so many times that my father's nightmares as a child were about not being able to find his mother and siblings.

In contrast, I thought the proposed move to Boulder was a terrible idea! Our differences were about distance, but once my father discovered his federal retirement income was exempt from state income taxes in Colorado, there was no stopping him.

Speaking very firmly, I said, "Daddy, I don't want to be helped in this way. Also, I really want to limit Olivia's contact with Mom."

His response, "You always want to have things your way, don't you?"

We were stalemated in mutual positions of projected incompetency. He thought I couldn't manage being a single working mother. I thought he couldn't manage my mother.

While my mother was still hospitalized, my father rented an apartment in a new building in northeast Boulder, flew to Phoenix, and returned to Boulder driving a U-Haul truck filled with their household furnishings, towing the Beetle behind.

Helpless to stop my father's plan to help me, I soon saw that the unspoken sub-text was that my father needed my help with my mother. He was exhausted from the struggle to keep her medicated and hated the obvious damage the psychotropic drugs inflicted on her body. The mental health system in Arizona was in greater disarray than Colorado's. Never good at managing the diseased brain of his wife, his efforts to do so had grown increasingly impotent.

Olivia with Grandpa Frank at the Boulder apartment of her grandparents, where there was a television, something she did not have at home in October 1971

Almost immediately after moving to Boulder, my father became a wonderful grandfather to Olivia—reading to her, helping her with schoolwork, watching her favorite television programs with her, and serving as his granddaughter's personal science tutor and chauffeur. With respect to his wife, however, my father continued to suffer from a paralyzing powerlessness. Nonetheless, to his great credit, he never abandoned her.

Over the next twelve years, it was necessary for me to hospitalize my mother sixteen times. The first commitments were at Boulder Community Hospital and St. Anthony's Hospital in Westminster. The last six commitments were at Bethesda Medical Center in Denver. She went through three psychiatrists, multiple drug regimes, and electroconvulsive therapy. Most of my emotional energy was spent trying to limit Olivia's exposure to my mother whenever she was off her medications or in a particularly active phase of paranoia. I felt measurable relief from worry about Olivia when an injectable form of Haldol became available that was effective for several weeks.

As chance would have it, just as I began the struggle to keep a protective distance between Olivia and my mother's demons, Dick Alton resurfaced. He and his second wife Linda had joined the staff of the Institute for Cultural Affairs, an ecumenical organization focused on self-help development in third world countries. Florence wrote to let me know Dick and Linda were now on a one-year assignment with the Denver branch. Six months passed before Dick called to ask if he could see Olivia.

Dick awkwardly entered the Marine Street cottage. After a brief period of small talk, Dick asked to see Olivia's room and glanced through her books. A telephone plan had been made for them to have a spaghetti dinner together at the Gondolier Restaurant, then located at the juncture of Marine Street and Broadway, three blocks east of the cottage.

As the biologically-speaking-only-father-daughter pair walked away, Olivia seemed frozen. I had tried to prepare my daughter for this paternal reunion, but it was impossible to hide all of my ambivalence. Further, it had been almost five years since Olivia had seen her biological father. Without memories—even paper ones—Dick was a total stranger.

While Olivia and Dick were eating dinner at the Gondolier, my mother began pounding on my front door—screaming in distress, while holding a branch from the silver maple that grew in the parking strip. At the sight of her holding the branch, I pushed past her, leaned over the front porch railing, and vomited into the Spirea bush. While I was retching, my mother continued screaming about being followed. Adamant that she couldn't tell me who was following her, my mother kept repeating that she had to hide inside my house. As usual, whoever it was wanted her secret papers.

Frank and Margaret on the twin Dinwoody's couch, Marine Street Cottage, June 1971

"Mom," I begged, "Please put the stick down!"

"No, I can't! It's my only protection!" she screamed.

My mother cowered in the bathroom with her branch, while I telephoned my father, who said he had been meaning to call me because my mother had been missing since the previous day. She had left their apartment telling him that she was "going shopping in Westminster"—her default destination. I asked him to come stay with Olivia while I took his wife to the hospital. After their move to Boulder, I had managed my mother's hospital commitments and psychiatric outpatient care without consulting my father. Her federal health insurance card was always in my wallet.

My father arrived just before Olivia and Dick returned from their brief dining experience. Since Dick needed to return to Denver, I asked him if he would help me drive my mother to St. Anthony Hospital. This was the second time I had asked him to be my mother's ambulance driver; the first was in Tucson in 1962.

Although far too much of my parenting involved the use of sugar, I arranged for my father to take Olivia out for ice cream so that she wouldn't witness the extraction process, which was awful. It took the full strength of both Dick and me to separate my mother from her branch, and to get her out of the bathroom into the back seat of my Beetle. Dick drove fast, as if he were an experienced ambulance driver, while I sat in the backseat with my mother trying to keep her from jumping out of the Beetle.

Perhaps understandably, it was three months before Dick made contact again. This time, he asked if Olivia could visit Linda and him at the Ecumenical Institute. When I told Olivia what Dick had proposed, she enthusiastically agreed. She fussed over what to wear and put together a bouquet of flowers from our garden for the stepmother she had never met. I drove her to Denver because Dick had not offered to provide transportation either way. It also provided an opportunity for me to do some up-to-date research on schizophrenia in the library of CU's Medical School.

As I shook Linda's hand, I thought she bore a strong physical resemblance to Florence, giving me a hopeful feeling for the relationship between Olivia and her stepmother. But when I returned three hours later, Olivia's facial expression and body language told me that whatever had happened was similar to her having spent another day without gold stars in Troy's kindergarten.

Olivia's painful disappointment was not a result of Linda. Rather Olivia was crushed because her long-lost-now-found father had kept her at too great a distance. Dick had introduced Olivia to each of the two dozen other Ecumenical Institute staff members with whom he and Linda lived without once acknowledging Olivia was his daughter. Her seven-year-old heart could not hold this inexplicable anonymity.

That night I introduced Olivia to my special remedy for the blues—wild dancing and sugar. We danced through our modest collection of records, skipping the too-mournful Leonard Cohen. In between songs, we ate a batch of homemade hot fudge with two spoons while it was still warm in the saucepan.

After that visit with Dick, all of Olivia's long-suppressed feelings of paternal abandonment surfaced. At first she struggled to quell them with food, just as I had taught her on the day her biological father broke her heart. I knew of no way to repair her paternal wounds. The distance between hurt and help was enormous, complicated by my unshakable guilt for having made choices that led to Olivia's loss of both Dick and Lary.

Since the expiration of the pre-nuptial agreement I made with my father in Berkeley and the end of Dick's co-habitation with Olivia and me, I had never reminded Dick of his parental obligations. Nor had I sought more active parenting on his part by reminding him of the importance of continuity in contact or mentioned his long-forgotten court-ordered financial responsibilities.

If money is symbolically filthy lucre, my unconscious had absorbed my mother's projection. For me, all financial transactions are painful. Paying for everything reduces my anxiety, as if I am perpetually getting rid of the suffocating snakes in my body. My character flaws kept me from reminding Dick of Olivia's need for him to have a role in her life.

As for Dick, my only explanation is that Olivia represented a shameful incongruity between his self-image as a devoted Christian versus a man who had forgotten his daughter. Olivia was the victim of her mother's guilt and her father's shame—both of which combined to create too much distance for a vulnerable young child.

-49-

Mad Manx Leads Me to True Love

Bank Street College of Education in New York City had created a model Head Start curriculum. After receiving a grant to test its efficacy in Boulder County, Head Start officials in Washington began asking me to consult with other Head Start programs. While training other Head Start directors in the implementation of the Bank Street model in New York City, I was introduced to Siobhan Oppenheimer, a program officer at the Ford Foundation. Siobhan invited me to dinner to discuss education of children for whom English was a second language—a dinner discussion that was the beginning of a friendship.

In June 1971, Siobhan called to ask for my help with a new Ford Foundation grantee, the Native American Rights Fund (NARF), which was about to open its headquarters in Boulder. Ford had given NARF a grant of one million dollars to establish a non-profit law firm for Indian tribes. The goal was to provide Native American tribes with resources similar to those of the National Association for the Advancement of Colored People and the Mexican American Legal Defense and Education Fund.

Siobhan began by telling me she had just learned NARF's founding director was to be a young lawyer named David H. Getches.

"Joan, I've seen this movie before! Getches has no management experience. He was president of his college fraternity and apparently believes that gives him the necessary experience. The person we intended to take the leadership role didn't want to stop working as a litigator. Also, he at least had the intelligence to know that he was not a skilled administrator. Somehow Getches volunteered to take his place. Would you be available to provide the managerial startup help that Getches is going to need, even if he doesn't know it?"

Although Head Start was winding down for the summer, I told Siobhan I didn't want another shotgun relationship—not even a work-related one. Siobhan promised there would be no Ford Foundation interference—she would simply give David H. Getches my contact information.

NARF had secured a lease, with an option to purchase, an old fraternity house at the northwestern boundary of the University of Colorado (CU) campus. The interior of the building revealed some of the reasons the fraternity had been banned, including empty beer cans, abandoned jock straps, women's panties, broken doors and dishes—the detritus of men gone wild.

David was tall, thin, and had a handsome angular face, but he seemed preoccupied during our initial meeting. Looking at my resume, he said that he had other candidates to interview. I left the NARF building on Broadway expecting to never hear from him again.

Two days later, David called, saying he would like me to be his "executive assistant"—offering an extremely low salary. Minutes after David's call, Siobhan called. When I told her I was going to turn down the offer, she asked me to postpone my decision for twenty-four hours. Before the day was over, Siobhan had arranged for a Ford Fellowship to supplement my salary, successfully pleading with me to accept David's offer.

David had been an Eagle Scout and had grown up with a mother who fully expected him to become President of the United States. David had a wonderful sense of humor, and, given his remarkable skills as a thespian, he might have been a better President than Ronald Reagan. David was particularly adept at playing "the Church Lady" the part first perfected by Dana Carvey on *Saturday Night Live*.

NARF, like Head Start, was another startup operation requiring long hours, seven days a week. Staffed by an idealistic group of young lawyers, including three of the first Native Americans to graduate from law school. NARF's mandate was to use the judicial system and other institutions to preserve Native American sovereignty, culture, and natural resources. Building NARF from the ground up was complex and compelling work. It was necessary for me to be firing on all eight cylinders and work long hours. Fortunately, within a few weeks I was able to form a successful professional partnership with David and we soon became close friends.

It was also time for Olivia to experience new challenges. Ready to start third grade and having outgrown Upland School, I enrolled her at Flatirons Elementary School, a neighborhood public school located three blocks from our Marine Street cottage and four blocks from NARF. Upland had seventeen students, Flatirons had three hundred seventeen.

Flatirons was close enough to our cottage that Olivia could come and go on her own as a latchkey child. Grandpa Frank frequently met Olivia after school to help prepare a snack, answer homework questions, or do a few housekeeping chores while she played with a friend. If Olivia had a dance, piano, or horseback riding lesson, my father was her loyal chauffeur.

The author and Olivia at the home of David H. and Ann Marks Getches, on Twin Sisters Road, Boulder County, November 1971

Olivia had been at Flatirons several months when my mother began secretly following her. When Olivia saw her Grandma Margaret, she ran to her Beetle.

Grandma Margaret put her finger over her lips and said, "You mustn't tell him that I know you are in danger! Don't let him know you have seen me!" before she sped away.

As was our rule, Olivia called me just as soon as she arrived home, telling me, "Mom, Grandma Margaret is following me!"

That night, I sat down with Olivia to try to explain the nature of the invisible disease inside her grandmother's brain, struggling to carefully modulate some of the current theories. It was a hard moment to do so because I had just read that Dr. John Nathaniel Rosen, (author of "The Perverse Mother"), had been named "Man of the Year" by the American Academy of Psychotherapy for his theories on "schizophrenogenic mothers" and the treatment of schizophrenia with what he called "direct analytic therapy."

That night I was sleepless thinking of Olivia and my mother, but in the morning, the onslaught of daily tasks displaced my oppressive fear of mental disintegration. Overwhelming work obligations pushed my anxiety about my future into the background, where it was reinforced by human denial, and the comforting assurances of Dr. Maurice Fox and Kenwood Bartelme.

In February 1972, eight months into my tenure at NARF, Robert Stuart Pelcyger (always known to his friends and colleagues as Bob) asked me to have dinner with him. It was a day when we were both working on NARF business in Washington, DC. Bob had curly brown hair and his hazel eyes seemed permanently crinkled with laughter. Married with two young children, Bob's innovative approaches to the legal problems of his Native American clients in Southern California and Nevada had been the primary impetus for the Ford Foundation grant. As a result Bob held a position of intellectual leadership at NARF.

After dinner, Bob and I continued our conversation in the deserted dining room of the Tabard Inn on N Street, across the street from NARF's Washington Office. I had known nothing about Bob's history or personal interests until that night. Born in Brooklyn, his family had moved to Valley Stream on Long Island when he was eleven. He had one sibling, a younger brother, Joel, who was in the midst of opening P.S. 1, a private pluralistic elementary school, in Santa Monica with Ellie Cobin, a parent.

At the University of Rochester, Bob had been student body president and studied with Norman O. Brown,[38] with whom he still corresponded. I had loved Brown's books, *Life Against Death* and *Love's Body*. Bob seemed extremely surprised when he realized that I had read Brown's books.

Bob had chosen Yale Law School over Harvard because Yale's flexible curriculum permitted him to earn law school credit for course work in other departments. After he finished his law degree, Bob had been awarded a Fulbright Fellowship. He had taken his wife Deborah and their newborn daughter Gale to England where he undertook a year of study with Owen Barfield, an English solicitor. Barfield was also a philosopher and poet, who wrote about the evolution of consciousness.[39]

That night Bob and I spoke about our favorite poets and our sharply contrasting childhoods. At three in the morning, trying to be as respectful as possible while stifling a yawn, I told him that I had an early morning meeting.

"Bob, it has been a lovely evening, but I'm exhausted. Is there anything else you want to discuss?" I asked.

Bob hesitated before responding, "Well, yes, there is. I wanted to ask if you would sleep with me."

Bob Pelcyger, 1972

I don't know why I agreed, except that Bob's overture was so charmingly naïve and it was the early 1970s when sexual mores were so relaxed such encounters were almost non-events. I think now that I granted his request because I was already falling in love with Bob's mind. While I remember laughing at his modest blue flannel pajamas, the sense of spiritual connection I felt with Bob Pelcyger disturbed my equilibrium. From that moment on, I felt preoccupied by his presence.

During our dinner discussion in Washington, I had told Bob that I had never heard of Owen Barfield. When I returned from Washington, I found a copy of Barfield's book, *"Saving the Appearances"* on my desk. He had written a note on the paper sack from the Colorado Bookstore that read *"If you have the hubris."* I felt annoyed because I had to look up the meaning of "hubris."

It was not long before other NARF staff knew about our encounter. Bob's wife, Deborah, and several other NARF wives gathered to give me the same careful and thorough dissection I had once given Elizabeth Budge in Logan and my father had given Elizabeth Taylor when Eddie Fisher betrayed Debbie Reynolds. The NARF wives held me fully responsible for Bob's marital betrayal. It was impossible to shift even a small amount of sinful blame off my platform sandals onto Bob's scuffed loafers.

Although Bob and I have now been married for forty-two years, I still feel shame for the pain our betrayal caused his first wife Deborah and their children, Gale and Jordan. Early in our relationship, I was convinced Bob and I would simply recreate whatever it was he was trying to escape. One day while Bob was dissolving his marriage to Deborah, I packed his bags and drove them to Pine Brook Hills, depositing them on the steps of the family he had abandoned. At that moment, I believed our ways of relating romantically do not change much from one partner to the next.

What I believe now is that partners must constantly work to maintain a mutually satisfying distance. Marriages fail when what is a comfortable and safe distance for one turns out to be too close or too far for the other—when a newborn child, a challenging job, a relocation, or a life-threatening illness increases pressure on existing marital boundaries. Most adulterous affairs seem to not be so much about sex, as about having a secret—a secret that changes the distance between self and other.

When Deborah's parents flew to Boulder to try to save their daughter's marriage, Bob went to meet with his in laws at their motel, where they presented strong arguments for trying to save the union.

Bob's only defense was, "You forget, I am in love with Joan."

But like every human being, I was vulnerable to betrayal. After Bob's divorce, but before we were married, he slept with another woman. He has always traveled a great deal as a part of his legal representation of Native American tribes, but however far apart we were geographically, we developed a ritual of saying goodnight on the telephone before going to sleep.

One night when I called, I immediately knew Bob was in bed with someone else. Bob denied my instincts that night and for weeks afterwards. The sexual betrayal hurt, but it was his deceit that changed the distance between us. Until Bob confessed, the distance between us was an unbridgeable chasm of distrust.

After Deborah filed for divorce, she took Gale and Jordan to Boston, along with all of their household goods, with the exception of three items: Bob's desk, his chair, and a large framed reproduction of Hieronymus Bosch's fifteenth century triptych, *The Garden of Earthly Delights*. Norman O. Brown had introduced Bob to Bosch's paintings and this was his favorite. His desk and chair went to NARF since they dwarfed the interior of the Marine Street cottage, but we hung *The Garden of Earthly Delights* on the east wall of the tiny living room. Bob made me feel as if I had just stepped into the Bosch's heaven. I loved the painting and the man who brought it into my life.

Olivia, who had previously met Bob at NARF, told me, "Bob has the nicest, kindest eyes."

Manx must have agreed with Olivia. On his first visit to our cottage, Bob entered and sat down on the twin couch. Manx immediately began rubbing against his polyester bellbottom slacks so Bob reached down to pet Manx. Within fifteen seconds, Manx was in his lap, having morphed into feline Silly Putty. Olivia and I could hardly believe our eyes!

Manx adored Bob from the moment Bob stepped into Manx's lair. Yet, Manx was a cat who, like my mother, could hardly stand to be touched, who always looked and behaved as if he expected to be betrayed or beaten, and who exhibited all the symptoms of feline paranoia. Manx represented a powerful piece of feline evidence that schizophrenia is not caused by "perverse mothering" because Grey Cat was a wonderful feline mother.

The most important contribution to the science of feline-human behavior Manx made was his use of cross-species intelligence to show me that Bob Pelcyger was an extraordinary man.

-50-

Henry Kissinger and "The Refrigerator Mother"

The reaction of my mother to Bob was the complete opposite of Manx, although she kept it secret for three years. She was on injectable Haldol* through the first two years, then running away from her demons in Canada and California through most of the third. When my mother returned to Boulder in 1974, Bob was living in the Marine Street cottage and she could no longer contain her paranoia.

The Watergate scandal was causing my mother's schizophrenic antennae to vibrate wildly. Her delusional mind had begun to link Bob with Henry Kissinger, then President Nixon's Secretary of State. My mother had become convinced that Kissinger was plotting to steal her secret papers. Bob bore a vague resemblance to Kissinger—they were both Jewish males, both highly intelligent; their faces were similar in shape, and each had curly hair. However, Henry had two decades and twice as many pounds on Bob.

On a sunny summer afternoon as I walked home from NARF, I could see that our refrigerator was sitting on the front lawn—door open, empty. My mother's Beetle was parked across the street. Picking up my pace, I entered the cottage in a state of alarm. My mother was sitting at the kitchen table; standing in the place of the old refrigerator was a new one.

Looking at the new refrigerator, but speaking to my mother, I said, "Mom, what's going on?"

---

* Haldol is the trade name for Haloperidol, an antipsychotic medication developed by Paul Janssen in 1958. The major advantage of treating schizophrenia with Haldol is that there is a long-acting formulation that can be given by injection every four weeks for patients who either forget or refuse to take the medication by mouth. The major disadvantage of Haldol is that it frequently results in a movement disorder known as *tardive dyskinesia* which may be permanent, as was the case for Margaret Lieberman.

My mother stood up and said, "I was hoping you would come home alone so I could warn you!"

Despair collected in my solar plexus. "Warn me about what?"

Henry Kissinger's "Poison Gas Refrigerator" in kitchen of Marine Street Cottage

She moved closer. "The poison gas! You must have been smelling it! Bob and Henry are trying to kill you by putting poison gas into your refrigerator! It is a slow, horrible death and Olivia could be killed too!" My mother's tone and demeanor sounded like the Wicked Witch of the West talking to Dorothy about her ruby slippers in the *Wizard of Oz*.

"Who is Henry?" I asked as a stall, trying to think of how to turn her mind around.

"Henry Kissinger! He and Bob are trying to kill you because you know about my papers. They may try the same thing with the new refrigerator too, so you will have to remain vigilant!" My mother was evangelically adamant.

Clearly Margaret Lieberman was off her meds. When she finally drove away, the old refrigerator was gone. She had hired a man to haul it away. As my mother stood facing him on the front lawn paying his fee, I could see from the look on the man's face she was warning him to be careful with what was a very lethal refrigerator.

Late that night Bob returned from Reno. After I explained why there was a new refrigerator in our kitchen, Bob brushed away my mother's paranoid projections. But I found them worrisome. For one thing, this was Bob's first

306

known inclusion in my mother's collection of "dangerous enemies"—as well as his first direct exposure to the complexity of my mother's paranoid thought processes. I saw the incident as yet another reminder that Bob shouldn't try to build a life with me.

Bob had already seen me through some scary lenses. I had tremendous difficulty trying to maintain an *optimal distance* between us. It was not only the challenges of living together with regular physical intimacy, it was all the other carefully forged boundaries I had drawn around my life in an effort to protect Olivia; I did not want another man in her life to disappear.

In addition, I had begun to feel incapable of creating and sustaining a blended family. I first met Bob's daughter, Gale, in the halls of NARF, and found her to be a charming and appealing five-year-old. With her curly brown hair and twinkly eyes, she seemed to have just stepped out of a Shirley Temple movie. A year passed before Gale and I met again under very different circumstances.

Olivia, Gale, and Jordan on the twin couch, wearing Margaret Lieberman's gifts of matching Kmart sweaters, Christmas 1972.

Gale and Bob's three-year-old son, Jordan, arrived from Boston for their first solo visit after Deborah had moved herself and the children east. It was a hot day in July. Olivia, Gale, and Jordan asked permission to walk four blocks to Steele's Market to purchase Popsicles. They waited impatiently while I went inside to get my purse. Then I made my first mistake as a future stepmother.

Gale and Jordan were not yet my stepchildren, but as an only child, having raised an only child, I had below average sensitivity to the dynamics of sibling rivalry. Without forethought or malice, I placed the money for the three Popsicles into Olivia's pocket, the eldest and only knowledgeable neighborhood guide amongst the threesome.

Gale immediately began punching, kicking, and biting me while screaming, "I hate you! I hate you!"

I don't remember how I managed to get her into the house and onto her bed for a time out, all I remember are the bite marks on my arms. That night, as I lay in another bed next to her father, I told myself that Gale's rage contained all the pain packed into her psyche by the separation and divorce of her parents, as well as by her demotion from eldest to middle child. Something about Gale's sudden attack triggered memories of my mother. I still feel vulnerable and afraid whenever someone makes a sudden move from normal to angry outrage.

In the midst of these conflicts, my body spoke up. My neck had been a source of discomfort since the truck accident, complicated by my triple-shift life—trying to create a loving blended family while carrying complicated professional work responsibilities, and last, but not least also managing my mother's medical care.

Boulder was already on the cutting edge of various atypical health trends. A friend suggested I get "Rolfed" as a way to ease the neck pain. Rolfing is a deep tissue technique meant to separate the fascia from the muscle so that the muscles and the skeleton return to natural positions. Being Rolfed was consistent with the reminder of Mildred Verhaag to "Stand up and tuck in!"

Unaware of any risks, I went to my first Rolfing session in the office of Emmett Hutchins, one of the early disciples of Ida Rolf,[40] the creator of the Rolfing technique. Being Rolfed meant laying down on a table to have my muscles worked in a way that woke up every nerve sheath in my body. The work was done over ten sessions, each about a week apart. At first, after a day or two of feeling sore and teary, I did feel remarkably better, taller, and was standing straighter.

Emmett Hutchins could not complete my last two Rolfing sessions. Instead, he sent me to another male Rolfer. The last session is focused on the head, particularly the face. When the substitute Rolfer, who had a disagreeable body odor, put on rubber gloves and started working in my mouth and nose, I was lost.

Rolfing is now advertised as a technique capable of releasing body memories that can be emotionally cleansing. I don't recall how it was being advertised in 1973. All I know is that I had a full-blown panic attack—my body became hypothermic and I was shaking so hard I could hardly drive home. I was treated in the emergency room and then admitted to Boulder's only psychiatric ward on a twenty-four-hour hold.

Upon discharge, I went directly to the office of Dr. James E. Marquardt. He had provided support to Deborah as she sought to recover from the pain caused by Bob's adulterous betrayal. Bob had bravely asked Dr. Marquardt to accept the source of Deborah's pain as a patient since he was no longer treating her.

During my initial meetings with Dr. Marquardt, I was only able to communicate by drawing images of my dreams. They were strange shower scenes with stick figures, although I drew myself as a worm. I spent the next decade in therapy with Dr. Marquardt. Having served as a physician in Vietnam, he had first hand understanding of how old trauma remains embedded in the body until accidentally triggered.

Several weeks passed before I was able to go back to work at NARF. While working at home I wrote grant proposals, reports, and NARF's newsletter, but my most successful work product was sewing an apricot cashmere cape for Olivia to wear when she accompanied Bob to the NARF Christmas party. Bob's love, kindness, and steadfastness throughout my recovery, plus Olivia's increasing reliance on him, became important building blocks for our shared future.

Bob and I were married in the backyard of the Marine Street cottage on August 23, 1975. David H. Getches performed the ceremony, having paid, like so many others, one dollar for a certificate that made him a minister in the Universal Life Church.

Ten days before the wedding, I finally managed to acquire a real wedding dress, a sample from Neusteter's Department Store in Denver, which the bridal sales staff promptly altered and delivered. The floor length gown was white linen and had a gently scooped neckline trimmed in a band of heavy cotton lace, not rickrack, but similar. I believe Mildred Verhaag would have given me her stamp of approval. Although ineligible for a veil, I was also sans girdle.

Nonny and Bob's brother, Joel, were the matron of honor and best man. Olivia, who had just turned twelve, and Gale, age nine, wore their first long dresses, while Jordan, who was about to start kindergarten, insisted on carrying his new Superman school lunch box.

The author before her first real marriage ceremony, August 23, 1975

Both Bob's and my parents were present at the ceremony. My mother's dress was eerily similar in color and style to the one Commander Margaret bought at Macy's for me to wear while reciting my shotgun wedding vows to Dick Alton in Berkeley. Bob's parents, Ruth and Eugene, came from their new home, having moved from Valley Stream, Long Island, to a retirement community in Hollywood, Florida. Although they must have been close to heartbreak, they carried themselves with extreme dignity on our wedding day.

Bob's first marriage ceremony had been performed by a Rabbi on the roof of the St. Regis Hotel in New York City. Deborah had been raised in a wealthy Jewish family who then lived on Central Park West. The marriage had produced a perfect family, a girl and a boy. In a painful contrast, on the occasion of Bob's second marriage, Ruth and Eugene found themselves observing their first-born son being downgraded and demoted to what must have seemed like hippie Hell.

In Boulder, Bob and Deborah bought a new home in Boulder's exclusive Pine Brook Hills. Now, his second marriage was taking place in the backyard of a rundown rental cottage. Not only had their son left behind what appeared to be a perfect life and family, he was marrying a twice-divorced Gentile woman— a real *shiksa*.

A few close friends and relatives attended the actual ceremony. We only agreed to a reception after a shy, normally-almost-mute Oglala Sioux Indian named Oran LaPointe, who worked at NARF, said, "Oh, can I come?"

Oran LaPointe's excited interest in our upcoming marriage was so unexpected, so startling, we lost control of our plan. Our closest friends staged a reception as a wedding gift at the Four Mile Canyon home of Bruce Greene and Susan Andre. Guests danced to the electric piano music of a cantor, who also sang popular songs. There was an unbelievably delicious *salmon en croute* made by Ann Marks Getches and two different wedding cakes—one vanilla with raspberry filling, the second, chocolate with fudge frosting.

Oran LaPointe never showed.

Just after the marriage of author and Bob Pelcyger, Marine Street Cottage, Boulder, August 23, 1975. Left to right: Nonny Thomas, Joel Pelcyger, Gale, Olivia, Bob, Jordan, author, Frank V. and Margaret Lieberman, Eugene and Ruth Pelcyger.

Bob and I took a week-long honeymoon in San Diego. Now familiar with the length of Bob's bathroom breaks, I had begun to imagine giant outputs. My own trauma-induced, highly protective habits were to get in and out of any bathroom in less than one minute, so I asked to see one of Bob's bowel movements. This was a pretty invasive request—definitely not appropriate distance for anyone over three years of age.

Bob's response to my request was, "Only on our wedding night."

The day after our wedding, we flew to San Diego and checked into a hotel on Shelter Island, where I completely forgot about my pre-marital request.

Bob and I began calling my mother's replacement refrigerator "Henry"—one of many efforts we made to maintain a sense of humor about my mother's mind. After our marriage, we bought the Marine Street cottage and undertook a major remodel to better accommodate our blended family. Henry was moved to the basement to handle overflow from a new fancy double door model with an automatic ice maker.

Marine Street Cottage after 1976 remodel guarded by Grey Cat (left) and Manx (right).

I never open Henry, or any other refrigerator, without thinking of my mother and the mistaken psychiatric theorists, who in trying to find a metaphor for what they assumed was a lack of maternal warmth in the mothers of schizophrenics, coined the term "Refrigerator Mother."[41]

-51-

Work and Distance

Work can be a burden or a joy, servitude or a source of self-actualization, but for most of us it is something in between. Whatever else it is, work is also an essential distancing device—one that is an important aspect of my family history.[42]

The childhood of Edward Robert Beck (my Pepa and maternal grandfather) was suffused with pioneer sod-breaking. Mud made him dream of becoming a gentleman, of trading his caked overalls and boots for a fine suit, clean shirt, and silk tie. Eventually, he learned to market his appreciation for the finer things by selling automobiles to other dreamers in the far corners of the West. One thing is certain about my maternal grandfather's work—it kept him at a considerable distance from his wife and five children.

In September 1919, just past his eighth birthday my father caught a final momentary glimpse of his own father under a streetlight in Ogden, Utah. Jacob Liebermann had been away for many months, serving in the Army of Occupation in France, disguised as a single man with no dependents. When the Army learned otherwise, Jacob was sent home, where he punched holes in a billboard so he could surreptitiously surveil the family he had abandoned. Seeing Hockey Moulding, a young man who had befriended his family, leave the house of his wife and children enraged him. A short block away Jacob ran up behind Hockey and stabbed him in the back.[*] Within the hour, Jacob Liebermann, then thirty-three, hopped a freight train headed east. He never saw his wife, or any of his four young children again.

When my mother and father fell in love, they were two father-starved children whose marriage mirrored paternal absence. I do not know when my mother

---

[*] No police report was filed to protect Anna Tuite Liebermann from further humiliation.

realized that her husband's work was going to keep him away from home as much as it had her father, but it did.

Work also created considerable physical distance in my marriage to Bob. For the first six months after our wedding, Bob was away from home, working in Washington providing representation to five bands of Southern California Indians before the Federal Power Commission. That long absence set the tone. After our marriage, I left NARF and began working as a management consultant—work that provided me with my own kind of *optimal distance*. I had thirty-four years of experience as a diagnostician. My survival depended on constant awareness of my mother's mental state. I refined those skills as a specialist in organizational diagnosis and leadership behavior. Thus, for Bob and me, our cohabitation began part-time and stayed that way.

My work as a management consultant also provided a safe space away from my mother. Between 1970 and 1982, my mother's madness eventually embraced most problems in our culture, plus a few extras: the devaluation of paper currencies; world economic crises; governmental secretiveness; the omnipresent FBI (which was either following her or trying to recruit her); biblical mysteries; the return of Christ; the abuse of children; the persecution of women, minorities, Jews; Mormon Church corruption; and her persistent paranoia about poisoned water supplies and poisonous snakes.

Blended Hanukkah and Christmas Celebration, Marine Street, December 1981. Left to right: Gale, Bob, and Jordan Pelcyger, Frank and Margaret Lieberman, and Ruth Pelcyger

On dozens of occasions during the last decade of my mother's life, she would announce she was going to the Westminster Mall. More often she went elsewhere. She would be gone until after dark, or overnight, or for days, sometimes for weeks, and on three occasions for months. When my mother first began disappearing, my father and I would file missing person reports and spend long hours looking for her. Eventually we stopped searching or alerting law enforcement. Trying to find "Missing Margaret" became, as my father often said, "A waste of taxpayer resources, as well as a waste of our lives."

Most families eventually let go of a loved one suffering from the uncontrollable impulses of schizophrenia. The awful futility and the almost universal denial by schizophrenics that they are suffering from a mental illness, magnified by inadequate mental health services, are why so many schizophrenics are homeless, wandering the streets, and too frequently taken to jail. The men, both young and old, are disguised by long hair and beards, while archetypal schizophrenic women can be seen bundled up, even in warm weather, with shopping carts filled with odd assortments of personal property. Instead of pushing a shopping cart, my mother's foot pushed on the gas pedal of her Beetle.

Soon after my parents moved to Boulder, my father divided their assets. Every month he wrote my mother a check for half of his federal retirement pay, while continuing to pay their household expenses. He soon refused to listen to her worries about money or impending economic disaster. Using the standard relocation remedy, he moved her six times in Boulder in search of relief, while simultaneously attempting to stay close to Olivia and me. They used twin keys to open the front doors of their shared shelters, but were separated in all other respects.

Among my mother's first long distance escapades were trips to Canada where she stored gold and silver coins in various safe deposit boxes in anticipation of the failure of all American banks. Those coins are still somewhere in Canada. Margaret had at least three bank accounts in Switzerland when she died, each with less than five thousand dollars. She trekked to Oklahoma to check out the oilfields related to her penny stock purchases; to Arizona to be close enough to Mexico to escape the FBI; to California for the same reason; and finally to Utah because she was convinced the FBI would never think to look for her there because they knew she hated Mormon men. She put over three hundred thousand miles on two different Beetles. Each was always filled with boxes of her papers, books, and multiple gallons of Knudson's apple juice.

Whenever she was missing, I found it hard to sleep. Often I would spend the night driving around the Denver metropolitan area looking for her car. I never became immune to concern about her whereabouts, even though I was repeatedly inoculated by futile searching.

In 1977, my mother left for the Westminster Mall and didn't resurface for a month. When she finally called my father, she told him she was living in Tucson. She wanted him to send her a box of clothing. Several months later, her deteriorating condition caught the attention of her apartment manager. Police located my father from the address on my mother's rental check. My father flew to Tucson and drove her back to Boulder in her Beetle. Convinced the water in her apartment had been poisoned, my mother hadn't bathed or drunk water for several weeks. She survived because of her substantial supply of apple juice, perhaps a beneficial carryover from Brigham Young's stockpiling revelation.

In Boulder my mother was diagnosed with congestive heart failure by my father's cardiologist, Dr. Edward Turvey. For the next five years, she frequently had to be hospitalized for fluid overload and given intravenous Furosemide (Lasix) requiring urinary catheterization. After several days of treatment, she would be discharged twenty to thirty pounds lighter.

My mother's last escape (or escapade) ended in Salt Lake City in October 1980. Having indicated her intention to go to the Westminster Mall, she instead drove to Salt Lake City.

When she stopped for gas at Little America in Wyoming, she purchased a Heath Toffee Bar. Pulling back onto Interstate 80, she took a bite and broke a tooth. On the outskirts of Salt Lake City, she found a dentist to put a temporary cap on her tooth and ended up renting a trailer from him.

When it started to rain, the trailer roof leaked. She became convinced that the FBI had seeded the clouds with poison just for her, based on an electronic signal from the special chip the dentist had placed inside the temporary cap. Screaming in fear, she called her oldest sister Vermilla in Bountiful. Her sister called my father.

My father flew over the Rocky Mountains. As he drove her back to Boulder, my mother repeatedly saw imaginary wounded women opening the passenger door and jumping out of whatever car was on the highway in front of them. At her insistence, my father stopped more than a dozen times to search for the wounded women only my mother could see.

After my father brought her home to their Boulder apartment on Cedar Avenue, my mother put her demons to work full-time on the contents of the Dead Sea Scrolls, searching for secret scriptural messages. It was work that kept her safely preoccupied until the end of her life.

-52-

My Beloved Daughter

If you want to understand any woman, you must first ask about her mother and
then listen carefully. Stories about food show a strong connection.
Wistful silences demonstrate unfinished business.
The more a daughter knows about the details of her mother's life—
without flinching or whining—the stronger the daughter.

Anita Diamant, *The Red Tent*[*]

During Olivia's childhood, I over-practiced truth-telling, although I did not tell her the full panoply of fears I carried in the years before her birth or during her infancy. Nonetheless, Olivia was often swamped by the legacy of the secrets she inherited. She was eight when I first tried to explain the nature of schizophrenia.

Until then, all I had said was, "Grandma Margaret can't help worrying about you. She has an illness in her brain that no one seems to be able to cure."

I didn't need to tell her I was terrified of being taken over by the same phenomenon or what risks she also faced. Olivia's unconscious was already imprinted with those worries.

Whenever my mother began to focus on Olivia's safety, my anxiety about my own mothering increased. I had come to understand that the dangers my mother imagined were a psychological cover for her own harmful intentions. As I became more familiar with the characteristic patterns in paranoid schizophrenia, I found myself constrained. Like all parents, I felt a natural concern for Olivia, but my mother's constant fearmongering made it hard for

---

[*] Diamant, Anita. *The Red Tent*. New York: Wyatt Books, St. Martin's Press, 1997.

me to express worry or caution. Finding the appropriate distance between us as parent and child was a struggle.

One night, when Olivia was thirteen, she left home after dinner for a sleep-over at the nearby home of her best friend, Becky. At three o'clock in the morning, I suddenly woke and roused Bob.

"Olivia is not at Becky's. I know something is wrong!"

A call to Becky's mother soon confirmed that she thought her daughter was spending the night at our house. We discovered the girls had gone with a third friend to the CU apartment of her uncle to try a prohibited pharmaceutical. With the help of a Boulder police officer and Grandpa Frank, we found the three girls and brought them home.

Because I inherited thin psychological boundaries from my mother, one of my greatest challenges as Olivia's mother was deciding which information to act upon. It required constant second-guessing of my instincts.

Olivia continued to have a pen pal relationship with Grandma Florence, sustained by two visits Olivia made to the home of her paternal grandparents in Bakersfield when I traveled to California for Project Head Start. The summer Olivia turned nine, Florence and Harry drove to Boulder, with another granddaughter and took both to Missouri to visit relatives. Olivia was very excited about this adventure, but I knew that her hopes for the trip were too high. It was a long, hot drive for two squirming nine-year-old girls in competition for their grandmother's attention. She returned in a state of disappointed exhaustion. After that journey to Missouri, Dick and Linda moved to Denver, where Dick accidentally broke Olivia's heart by hiding his forced fatherhood from his Christian colleagues.

Bob wanted to adopt Olivia and, after we carefully presented this option to her, she agreed. Dick Alton relinquished his parental rights without comment six months before my marriage to Bob. Judge Horace Holmes presided over Bob's legal adoption of Olivia in Boulder County District Court a few weeks after our wedding. Before and since Bob has provided Olivia with the kind of paternal support every child needs and deserves.

Olivia was fourteen when she began working part-time at Fred's Candy Store, then owned by a NARF secretary and her husband. Olivia has a vivid memory of my having sent her out to knock on the door of every business on Pearl Street in search of work with her resume in hand. I only remember how excited she was when she came home that day with a sales position at Lawrence Covell, a high end Boulder clothing boutique.

As Olivia moved through her high school years, she blossomed into a beautiful young woman. It was more than her physical attributes; Olivia had a comfort with honesty that made a lasting impression on everyone she met. She struck a healthy balance between studies, extracurricular activities, and socializing. Elected Head Girl of Boulder High her senior year, she led the student body with hard work and humility. Olivia reminded me every day that she not only had survived her legacy, but had mastered it in a way that made her a woman of integrity and strength.

Olivia as a high school student 1978

Having lived in the West her whole life, Olivia wanted to go east for college. We toured her choices together (the University of Pennsylvania, Radcliffe, Sarah Lawrence, and Cornell) before Bob's enthusiasm for his educational experiences at the University of Rochester led her to choose his alma mater. But three days before Olivia was scheduled to leave for New York, she still had not begun to pack. We got past that ambivalence with a great deal of help from a highly experienced packer, her Grandpa Frank.

Olivia and I flew to New York City to shop for some critical items for her college wardrobe. It seemed as if Mildred Verhaag was the only model for how to mother Olivia at this juncture. After scouring the City for the perfect collegiate purse and raincoat, we flew to Rochester where the carefully packed boxes sent by her Grandpa were waiting in the mail room of her dormitory.

Olivia had been assigned to a dorm suite and arrived before her other roommates because she was attending the last freshman orientation session. When I said goodbye the next day, Olivia looked so forlorn, so alone, that my heart barely allowed me to get in the taxi. I might have stayed had I not been overdue in North Carolina where a group of lawyers were in revolt. Only that consulting commitment kept me at an appropriate distance.

Olivia successfully hid the depth of her homesickness until she returned to Boulder at Thanksgiving. She looked grey and thin, like someone released from a prison camp.

Her first words were for Bob, "You didn't tell me that Rochester is the second most overcast city in the country! Nobody smiles there!"

Bob's defense, "The seminars on the main quadrangle on sunny days must have been the only ones I remembered."

Olivia reluctantly got back on the plane after Thanksgiving break, and again after the month-long winter break, but twenty-four hours later she called from Rochester.

"I am coming home! I am eighteen years old! I already have my ticket. Don't even think about trying to stop me!"

Olivia managed to remain in Rochester, although her first year was harder than I could imagine. I had no experience with missing home.

By the end of her first year of studies, Olivia was more settled, having decided that she wanted to become a clinical psychologist. Her freshman year course work had exposed Olivia to some of the literature on the theories of schizophrenogenic mothers. The first brain studies using computed axial tomography (CAT scans), followed by magnetic resonance imaging (MRIs) were being used to examine the brains of people suffering from schizophrenia. On the first day home for the summer, Olivia was full of questions.

"What caused Grandma Margaret to have schizophrenia? Do you think I am going to develop it? I'm worried because the research seems to show a pretty strong genetic link. Most of the literature puts the blame on cold mothers." She waited for my answers.

Putting my arm around her shoulder, I took a deep breath, "No one knows exactly what causes schizophrenia. I understand your worry, but I also know that even among identical twins, the incidence is only fifty percent. But more importantly, I have heard of no other cases in Grandma Margaret's huge Mormon family and I am now long past the typical age of onset."

Olivia's worried query came two days before my fortieth birthday. While I felt safely out of the reach of schizophrenia, forty years of living with my mother had left many other scars.

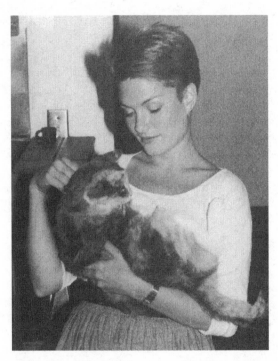

Olivia cuddling Grey Cat upon her return from Europe, August 1984

Holding Olivia close, I tried to sooth her, "I may have hurt you in many ways, but I promise none of them were deliberately malevolent. I understand your worry, but I believe whatever happened inside Grandma Margaret's mind is not going to happen in our family again. When I look at you, I see someone who is going to be an extraordinary therapist. I believe you will carry her painful legacy by using it to help others."

What I couldn't foresee at that moment was that Olivia would become much more than one of D.W. Winnicott's "good enough mothers." She would become an extraordinary mother to two delightful daughters, as well as a compassionate clinical psychologist.

But on that bright June day, after sharing a good cry together, we made a batch of hot fudge, eating it out of the pan with two spoons while sitting in the sun on the front steps watching Grey Cat give Manx a thorough tongue bath.

-53-

Death—the Ultimate Distance

The next day, my mother knocked on our front screen door. Switching on my guard persona, I descended the stairs from my upstairs office. On that warm June day, my mother was wearing her heavy navy blue raincoat, carrying her matching Kmart vinyl handbag and a nylon book bag filled with her papers.

A week earlier, we had celebrated her sixty-third birthday with dinner and a homemade cake—an emotionally draining effort. I had hoped for a longer interval between encounters. One of our front French doors was open so only the screen door separated us.

"Hi, Mom, what's up?" I asked.

My voice reflected an imperfect attempt to stifle the irritation I felt at her unplanned interruption, tinged with innate fear.

"Oh, it's nothing." She said. "I just wanted to see your face—I don't want to bother you. I'm sure you're busy working."

Since my mother seemed to be in one of her more functional states, I said, "Come in, Mom. It's hot out there. How about something to drink?"

She made her way through our entry into the kitchen, and up the single stair to the dining area and dropped into the nearest chair. In that short distance of about twenty feet she was breathless.

"Where is your portable oxygen tank?" I asked.

"Oh, it's in the car; I didn't plan to stay," she responded.

"It seems like you're uncomfortable. Why don't I go get it?" I spoke as I poured her a glass of fresh squeezed orange juice.

"No, I just need a minute more to catch my breath." She sipped the juice, marveling at how good it tasted.

I urged her for maybe the tenth time to buy the juice for herself, but she deflected the suggestion as an extravagance. Every once in a while I fell into the trap of trying to loosen my mother's Depression-formed spending habits. She asked for a refill, a rare instance of token acceptance on her part. Something softened in me as I moved into an unfamiliar zone. I decided to tell her about a pair of baby robins in a nest just outside my upstairs office.

Whenever my mother caught sight of a robin, she would exclaim, "There's a robin!"

She expressed no interest in any of God's other creatures, but for some reason robins were treated as a rare species.

"Mom, there are two babies and they are getting livelier every hour." I said, "The mother no longer puts up a fuss when I open the door for a closer look. You should come up to see them. You'll be able to get really close, the babies look like twin fluff balls and are so cute! I'm going to go out to the car and get your oxygen tank."

I knew the stairs would be hard for her, but I judged the avian close-up worth the effort.

My mother remained firm, "No, I really don't want to try it. I feel tired and I think I should go home."

As she spoke, the phone in my upstairs office began ringing. I excused myself for a moment. When I came back down the stairs, my mother had already slipped out, her Beetle no longer parked in front of our house. I went into the kitchen to clear her glass and found a note she had written on the paper napkin.

It read, "*I hope you will find the time to read this book. It is a very important book. XOXO Mom.*"

Beneath her napkin note was a book entitled *Angels on Assignment* by Charles and Frances Hunter as told by Roland Buck. The wife of Roland Buck had written the introduction—an indication no one else would. In the last two decades of her life, my mother had become a full-time missionary for the spiritual mysteries of the world. I took the book upstairs and put it inside my bedside cabinet. It was not the kind of book I wanted my clients to see on my office bookshelves, nor did I have any intention of reading it.

My mother died twelve hours later in the pre-dawn hours of my fortieth birthday. I have many powerful memories of her, but the most persistent physical image is of her body on her bed that morning.

When I was allowed to go into her bedroom, she was lying on her back, her head turned slightly toward a southern window. Her hair had been recently cut in a short Dutchman's bob, so she was not in the process of "recovering from" or "about to need" a home permanent. She was wearing one of her numerous floral Kmart nighties, sleeveless, knee length, with a modest bodice. The lower half of her body was covered by a floral sheet and an acrylic blanket.

My mother's left arm was bent so her hand touched a small earplug that ran from her left ear to an inexpensive tape machine—she had been listening to a charismatic preacher from Tucson, Arizona. Her bare right arm was extended; the hand open, palm up. The inside of her extended arm had a strange vulnerability, like the breast of a young animal. Her skin, heavily freckled, except the inside of her elbow, had the sheen of alabaster in the morning light. The only obvious indicator of illness was tubing running from her nostrils, across her cheeks, over her body to the oxygen tanks standing like twin missiles at the foot of her single bed.

My mother died peacefully in her sleep, so peacefully that my father had gone into her room three times trying to rouse her. Not until his third try did he realize that his wife of more than forty years was dead. For months afterwards my father was very hard on himself for this lapse in his scientific observation skills. Their marital relationship had reached such a distanced accommodation that they rarely looked directly at each other.

That morning, my father stood at the foot of his wife's bed, calling, "Margaret, Margaret" jiggling the mattress with his knee in a non-intimate attempt to wake her, while gazing out the window without looking at her.

My mother only slept deeply in the mornings. Her congested mind, as well as her congested heart, made sleeping at night hard. Because of this pattern, my father was not surprised when she failed to respond. He left her room twice hoping she would rouse herself. Only on his third attempt did his growing impatience cause him to look directly at his wife's face. Then he saw the discoloration around her mouth from the pressure of the oxygen tubes. Rigor mortis had begun.

My father called Dr. Turvey, whose office in the Boulder Medical Center was less than a block away. Unfortunately, it was Dr. Turvey's day off. His receptionist transferred the call to another doctor at the Medical Center and the second doctor's nurse told my father to call 911. After that it was out of my father's hands—what should have been private became a public scene.

The Boulder Fire Department sent a truck, its sirens screaming, to the front of their apartment building. The fire truck, carrying a paramedic team, was followed by a police car, also sounding a siren. Four firemen and two policemen entered and quickly made the obvious determination—my mother was dead.

When my father telephoned me with the news, I was gingerly drying my hair, my scalp still tender from a rare middle-of-the night headache. I woke Olivia and left a message for Bob at his law office. As Olivia and I pulled up in front of my parents' apartment, the firemen and paramedics were climbing back into their fire truck. Inside the building, a policeman stood outside the door to their apartment. His radio unit was broadcasting static police calls throughout the entry hall and into their modest living room. A second policeman stood at my mother's bedroom door and an anemic young woman sat stiffly in my father's chair. My father sat across from her in my mother's chair.

My father stood up and came toward me, looking his real age, seventy-three, rather than the "about fifty" presumed by strangers. As I put my arms around him, he began to sob, while telling me he had been trying to wake my mother so they could deliver my birthday gifts before Olivia and I left for a day-long hike in the Indian Peaks Wilderness west of Boulder.

Pointing to two packages in pink wrapping paper on the dining table, he kept repeating, "I wouldn't have tried to wake Margaret if it hadn't been your birthday."

It was as if he believed his wife might still be sleeping had he not gone into her room. I thought it was more likely that she would still be living if I had never been born.

After rubbing my father's back and trying to sooth him with words, I turned to go into the bedroom to see my mother's body. The policeman at her bedroom door stepped into my path.

"No one is permitted to go into the room until Detective Stone arrives and completes his investigation!" he said.

Stunned by the implications of his denial, I turned toward my father for answers, but he had his back to me as he looked out the front window as if ashamed to meet my eyes. The young woman still sitting in my father's chair was staring at the floor.

"Who are you?" I asked.

The policeman answered for her, "She is an intern with the department. She's here to observe."

The room seemed smaller than usual; the mute young woman, the policeman's bulk, his gun, the crackling radio, and the scent of cigarette smoke on his body and uniform were occupying a lot of space. I began to argue with the policeman about my right to see my mother. It was important to me and I wanted my daughter to join me. I didn't understand why we couldn't go into her room to say goodbye.

I felt as if my mother was getting further and further away from me every second that this man blocked my path. My entire existence had been focused on trying to keep my mother at a distance, yet at this moment I felt a mixture of panic, rage, and fear because I was being prevented from immediately going to her side. Neither my feelings, nor the behavior of the policeman, made sense.

"I'm sorry, but that's our procedure," he said.

His officious behavior was infuriating. I wanted him out of my parents' apartment. First, I asked that he turn down his radio volume. Then I asked that he turn it off. When he refused to comply with the second request, I asked that he wait in the hall with the other policeman.

"Alright, if you insist," he agreed, "But only if the front door continues to remain open."

"You don't trust me not to go in?" I prodded.

"I'm not saying that, Ma'am."

"Then what are you saying?"

"Nothing, Ma'am...just our procedure," he spoke as he moved slowly out the door into the hall, leaving his scent behind.

Olivia had begun to weep. When my father sat down again, Olivia sat on the arm of my mother's chair, her arms around her grandfather's neck, her cheek on the top of his head. The young woman in my father's chair continued to stare at the floor.

"I think internships provide important training experience, but you are infringing on our privacy. I would prefer that you observe from the hall with the others," I said.

The female intern stood, still mute, and looked toward the open door, as if needing permission to move. I went into the hall to tell the two policemen we would like the intern to train elsewhere. The officer just banished to the hall motioned her forward. Finally, we were alone with the door open—the watchful threesome continued their surveillance duties just beyond.

Soon other tenants in the building passed by the open door, curious and concerned. All were elderly and well-acquainted with my father because he was the resident manager. Through the living room window I could see and hear the police car—the motor was still running, lights flashing, radio calls audible.

An hour passed before Detective Stone arrived, accompanied by the Boulder County coroner and his assistant. They asked a few questions about my mother's health and took the name and number of Dr. Turvey before going into my mother's bedroom alone. Two minutes later their investigation was over.

Taking Olivia's hand, we slowly walked into my mother's sun-filled bedroom. There was a strange smell in her room, molecules from my mother's death, mixed with those of her morning guests. The card table near her bed held a plate of half eaten green grapes, her glasses, papers, and several open books— one about the Aegean Sea along with three reference books on the Dead Sea Scrolls. Every surface in the room was covered with stacks of her papers. Her research notes, written in shorthand, clearly had been examined by one of the authorities, an act of surveillance she had long feared.

My father had followed us. "She looks peaceful, don't you think?" he said, his voice filled with the anxiety that accompanies the death of a spouse whom one hasn't loved perfectly.

My mother looked more normal than she had in many decades, which was both reassuring and painful. Removing the earplug from my mother's left ear and the oxygen tubes from her nose, I kissed her forehead and stroked her hair. Then I put my hand on hers. I was afraid to do each of these things, but conscious of trying to model a lack of fear for Olivia. My daughter sat on the bed, touching her grandmother's leg through the bed covers. Later I regretted that I had not laid down beside my mother's body to feel the now permanent distance between us.

There is no normative behavior for most of us in the presence of death, but I felt a tremendous loss of control. The moment was split, as if created in my mother's psyche. The coroner and his assistant were waiting in the hall, making noises to show they were impatient. I felt part criminal, part victim. The authoritarian nightmare wouldn't end until they left and they wouldn't leave without her body. At that moment I felt as if I would never again be capable of an authoritative act, never able to cover up, clean up, or clear out the confusion caused by my mother's mind.

The officious men who had invaded our lives that morning were likely suspicious of us. It was not hard to conjure what might be in local police records. Or maybe they simply sensed our deep relief, our long wish for an end to my mother's madness. Had they been courageous enough or humane enough to look into our eyes, they would have understood how often the husband and daughter of the deceased had tried to provide a safe haven for this woman, even though there was no such place on earth.

When I finally signaled the coroner, my father left the room and her side for the last time. My arm around Olivia, I watched as the coroner and his assistant wrapped my mother's body in a sheet, lifting it off the bed onto their gurney. As they removed the blanket and top sheet, I noticed the blood had begun to pool in the corners of her body. I asked the coroner how long he thought she had been dead.

"About eight hours," he replied with authoritative confidence.

It was then eleven in the morning, almost eight hours since I had been awakened by a rare headache. I had gotten out of bed and taken several aspirin. Had my mother somehow been trying to reach me? Or I her?

My father and I stood in the hall with the coroner making an uninformed, proximity-driven decision to use Howe Mortuary, while the coroner's assistant and the policeman loaded my mother's body into their van. Then we were finally alone with our confusing, conflicted grief.

Olivia prepared tea and toast, while I removed my mother's bed linens, soiled from death's involuntary release, to preserve this small measure of dignity for her. Olivia took the bundle from my arms; returning from the laundry room, she reported the soiled linens smelled of apple juice. My father had begun looking through my mother's papers for her will, a document she mentioned frequently. He found it in her strongbox, along with her eclectic collection of mining stock certificates, one hundred silver dollars, and a dozen gold Krugerrands.

I read my mother's will aloud sitting on the floor; while my father, in his chair, and Olivia, in my mother's chair, sipped Earl Greyer tea and nibbled on his homemade whole wheat raisin bread. The last will and testament of my mother was full of the usual oblique references to God's will and the destiny of both Gentile peoples, Mormons and Jews. With respect to her assets, my mother threatened to come back to haunt my father if he ever attempted to sell any of her penny stocks.[*]

---

[*] Twenty years later, a Charles Schwab broker determined that all of Margaret Lieberman's penny stocks were not even worth a penny.

"Proceeds from the sale of her writings," I read, "were to be distributed equally among all of her brothers, sisters, nieces, and nephews."

My mother had never been paid a single penny for anything she wrote, beginning with Tide contest jingles in 1952 to her latest submissions on the Dead Sea Scrolls to *Christian Life*.

The biggest surprise in her will was her purchase of a burial plot in the Louisville City Cemetery.

She had written, "*The plot was chosen because I like the view of the Rocky Mountains from there,*" a strange choice for a woman who was terrified of heights; followed by an additional instruction, "*The grave is to remain unmarked.*"

I began laughing and joked, "The absence of a headstone is her plan for hiding from the FBI throughout eternity!"

My father kept repeating over and over: "But we always agreed that we would be cremated!" as if this was a topic of conversation they had on a weekly basis.

In the final paragraph of my mother's last testament were specific instructions for her burial. There was to be no funeral service. We would find a box on the top shelf of her closet containing the clothes in which she wished to be buried.

She also had included a request for me, "I want Carol to go to the mortuary after my body has been prepared to make certain that my hair is combed properly and my lipstick is on straight."

I laughed again and wondered aloud whom she was expecting to meet.

Sitting on the floor below my mother's empty chair, I felt ashamed for never having spoken to her about what she wanted. It had been five years since Dr. Turvey gently told her that her heart would probably not last more than two years. Privately, he told my father and me it would last less than one. In my family there were more than the normal conversational taboos about death. Hearing Dr. Turvey's prognosis, panic washed over me—that, as my mother's physical body deteriorated, her psyche would match it cell for cell. I did not want to nurse her terrible thoughts for a single moment.

The box of clothes my mother had prepared for her burial contained the same dress in her preferred ice blue color that she had worn to our wedding. There was also a slip, a clean bra, and unbelievably one of her standard panty girdles! Were these girdles her substitute for Temple Garments? Momentarily mystified by what looked like two cotton balls, I began sobbing when I realized they were actually rolled up golf socks to keep my mother's always cold feet warm.

When Bob arrived, we over-briefed him on the behavior of the crowd of officials without speaking of the primary event. Bob took Olivia home to recover and I drove my father to Howe Mortuary.

The bronze and steel caskets were too ominous for me, too expensive for my father. We finally selected an orthodox Jewish casket. It was all wood and hand pegged to insure complete disintegration into the earth. It was as close as we could come to their cremation pact given her will—plus it was a choice of container containing some spiritual irony.

My father and I continued to occupy ourselves with the ritual tasks, correcting each other as we provided the *Boulder Daily Camera* with obituary information. My father had the hardest task of calling her brothers and sisters in Salt Lake City, Bountiful, and Phoenix. Although she was the youngest of their quintet, she was the first to die.

After those painful calls, we drove east in silence to the Louisville City Cemetery. I was surprised to see that it was surrounded by an ornate wrought iron fence and filled with mature trees. On that June evening, as we stood beside my mother's plot, the air reverberated with the calls of robins. Hearing their songs, I told my father we should have a graveside service. He was hesitant. But I reminded him all services were for the survivors and that we needed to try to bring some kind of closure to the life of Margaret Audrey Beck Lieberman despite, or maybe because of, the dictums in her last testament.

Early the next afternoon, having seen my mother's obituary, a woman counselor from Boulder's Unity Church called to ask if she could meet with us. The day before my mother died, the counselor had spent several hours with her. Rather than postpone what we dreaded, I issued an impromptu invitation for her to come to my father's apartment.

The Unity counselor dutifully but delicately explained to us, "Margaret was worried about a great many things in this world."

I could see that the counselor was deeply disappointed that we didn't ask her to elaborate. However, it would not have been possible to persuade her that she had glimpsed only a tiny portion of the worrisome problems stored in my mother's mind.

After the Unity counselor departed, I suggested that my father take a nap while I drove to Howe Mortuary. It was time to honor my mother's final request to check her mortuary makeup and hairdo.

Alone in the eerie entry hall, I let myself feel the fear in my body, while a Howe mortician, who closely resembled "Lurch", the manservant played by Ted Cassidy in the 1964 ABC television series, *The Addams Family*, went to retrieve my mother's body. Even in death, the prospect of seeing my mother brought the same physiological sensation as it had my entire life. Alerting every cell in my body, I placed an invisible emotional shield around my psyche, in an attempt to keep my mother at a safe distance. I had to behave with total rationality, devoid of all trust, not unlike the policeman who guarded her bedroom door.

As I stood to accompany Lurch, my knees buckled. Artfully taking my elbow, Lurch led me down the hall toward the viewing room.

Lowering his voice to barely audible, Lurch whispered, "The condition of the remains was such your mother's dress could not be zipped up the back, nor were we able to use all the undergarments you so thoughtfully provided."

As we entered the room where her orthodox coffin rested on a metal accordion Gurney, Lurch added, "The self-belt of her dress has simply been tucked around the waist area since it could not be buckled."

After I failed to respond, he said, "I'll give you some time alone with the remains."

My mother still looked serene, but distinctly different. I had never seen my mother's face and hands without freckles—the Howe mortician had used a very heavy hand with the foundation. She also now had very black eyebrows and an unsuccessful attempt had been made to draw lips on her face using day-glow orange lipstick. Clearly, her written request for my help with her mortuary make-up had been an inspired act of precognition!

I dampened my Aunt Mary hanky in the hall fountain, but in the end I had to use liquid soap and stiff brown paper towels from the restroom. The black on my mother's eyebrows was the hardest to remove and the orange lipstick smeared wildly on the waxy surface of her skin as I attempted to wipe it off. My hands were shaking so badly I faltered trying to make a straight line with her favorite Revlon Red lipstick on lips that seemed to have completely disappeared. I used my comb in an attempt to soften her bangs, but her hair remained stiff, like a bad wig.

Finished with my meager, mostly unsuccessful efforts at physical restoration, I pulled up a chair by the side of her coffin for a private debriefing. It was the first time I had ever spoken freely to my mother, having learned from the very earliest age to modulate all expression in her presence. As I grew older, I always switched to a private communication channel whenever I interacted with her, one where I was on high alert to her every nuance, while remaining neutral and distant in my own responses. My diagnoses were almost always right; the treatment plans less so.

My mother's presence was palpable. While I wished it was less so, maybe I was imaging intimacy since, while doing her mortuary makeup, I had just had more physical contact with her body than I had since I was ten when my mother got a blistering sunburn after a day spent floating in the Great Salt Lake. That day she had asked me to apply Unguentine ointment to her raw back, and it was one of the few times I remember having physical contact with her body.

"Oh, Mom," I said, and as soon as I spoke tears started spilling out of my eyes onto the muslin lining of her coffin. "I can't believe it's finally over. I thought I would feel relieved, but mostly I feel a sense of failure. I always wanted to cure you. I guess every child of a sick parent feels that way. Sadly, the harder I tried, the worse it got for you. I deeply regret my failure. Please forgive me for these last twenty years. I behaved more like your jailer than your daughter, which probably increased your isolation and loneliness. A lot of what I did was because I felt your paranoia was hurting Olivia. I didn't want her filled up with all your fears. I had learned how to defend against most of them, but Olivia was so vulnerable when you were going through your worst periods.

"In the last few moments of your life, I am hoping you saw a glimpse of something safer than you spent your life imagining. If there are spiritual beings, I ask them to give you back a sense of trust. I also have to tell you that I am relieved the mortician couldn't manage to put on your panty girdle. Your body deserves a break!

"I doubt you like the choices Daddy and I are making. We are doing our best, even though some things are different than you requested. I promise that I will try to take better care of Daddy than I did you. I see clearly now that I lacked the compassion needed to match my courage. While I always tried to remind myself that you were not able to stop the voices you heard, somehow I never managed to fully convince myself you had no control.

"I am still afraid that somehow your illness is lurking deep within my own brain. Olivia is afraid for herself in the same way. That's why I kept hoping you could trust a doctor, someone to whom your schizophrenia was not such a threat, someone who could really listen, without any fear. I am really sorry we never found such a person. I saw how you suffered and I backed away. I tried to save myself and Olivia from your incurable, mysterious disease. I hurt you most by my attempts to protect Olivia, but I had to, Mom. I had to!

"Look, there is one thing I want to ask of you. Between Utah, Mormonism, Daddy's atheism, and your mental illness, it has been difficult for me to explore my own spiritual beliefs. Would you try to give me some kind of sign if you find yourself in some kind of other reality? It would really mean a lot to me. I am surprised I am asking you for this. It just popped into my mind.

"So maybe I'll see you someday somewhere. I'll look for you and you look for me. I love you. That is the most unbelievable part of the whole terrible terrorizing pile of shambles. Something deep inside me never stopped loving you!"

The next morning during my mother's burial, there was a steady downpour, one missed by all the local meteorologists in their forecasts for the day. I did not imagine angels weeping, but rather saw the rain as a sign from my mother that she was really angry for our having ignored the "no services" instruction in her will.

My mother's sisters, their spouses, and her oldest brother had driven over the Rocky Mountains from Salt Lake in a single well-preserved Buick. Her youngest brother had flown from Phoenix without his wife. With her Mormon siblings and our friends, we huddled together under a too-small Howe Mortuary canopy; the eldest seated on metal folding chairs. My mother's casket looked small and vulnerable, suspended above the grave next to the excavated mound of soil discreetly covered with a sheet of plastic grass.

Despite the presence of the lady counselor from Unity Church, we had no spiritual leader. Instead, we clumsily worked our way through the program Olivia and I had prepared before dawn at Kinko's, a local print shop. It was a small memory book with photos of Margaret from birth to her last birthday party, just ten days past. Olivia and I had added a few of her poems (she had written thousands), as well as three of the scriptures she repeatedly invoked.

Only a few personal stories were told. Aunt Mary let the cat out of the bag by joking that while her family would miss Margaret, Kmart would really miss her! The truth was Kmart would miss my mother and her responsiveness to their Red-Light Specials, but anyone who knew her couldn't help but feel relieved by her death. To end the awkward gathering, simultaneously smoothing over our diverse beliefs or lack thereof, as well as our raw open wounds, everyone, except my father, awkwardly sang "Amazing Grace." Unable to either speak or sing, my father was blind with tears at the end.

The previous day, in search of physical finality, I had decided to fill my mother's grave, rather than tossing in token clods. Bob, Olivia, her friend Becky, and Art, Olivia's high school sweetheart, volunteered to assist me. Within minutes our good shoes and coats were covered in mud. I thought of Pepa.

My father, who had strongly disapproved of my plan, turned away, riding back to our house on Marine Street in the already crowded car of his wife's siblings for the traditional gathering.

By the time our work as self-designated grave fillers was finished, the rain had stopped. We drove back towards Boulder's Flatirons under a suddenly bright blue Colorado sky.

That afternoon, while we were crying, laughing, and eating with our guests on the garden deck, the two baby robins fledged. Watching them swoop down from their rafter nest to the cherry tree by Olivia's bedroom, coached by the calls of their anxious mother, I understood a new reality. Olivia, my beloved daughter, could now fly free, and so could I.

Robin fledglings

Optimal Distance, A Divided Life, *continues in Part Two.*

Acknowledgements

This narrative provides glimpses of the help I received from others—those still living and others now dead. Some gave me a smile, words of encouragement, a cup of warm milk, an infusion of hope, or a tender embrace. Others shared their recollections and insights in writing. Some recommended a book, or gave me the name of a connection. A few gave me courage by asking for my help. All were instrumental in my unexpectedly long survival and kept me from being lost.

Bread Loaf Writer's Conference provided a space for Blackie and Snowball to come back to life with the encouragement of Patricia Hampl, as well as Carol Houck Smith of W.W. Norton.

Distant librarians responded to my esoteric inquiries and both family members and my friends tolerated my long research process.

Scott S. Miller kept the wheels from flying off the complex task of preparing a two-part book for publication. In addition, he repeatedly provided emergency technical aid in the middle of the night and undertook the restoration of old family photographs.

T. Keith Harley's eye for the visual, combined with his exquisite design talents, brought Part One and Part Two of the manuscript to life on the covers and his designs for the interior pages.

Kim Brown Federici, Jennifer Breman, and Laura K. Hink were early readers who helped me to persist.

The corrections and comments of Charlotte Smokler, Ann Marks Getches, and Roland Evans were essential.

Above all, I am filled with gratitude for the remarkable editing talents of Emma Komlos-Hrobsky in New York City. Emma, in addition to being a wonderful writer, is an unusually careful reader. Her therapeutic insights and suggestions brought me closer to the core of my life.

Glossary of Individuals and Animals

* = *animal*

Alton, Dick. Author's classmate at EBHS and UC Berkeley; biological father of Olivia. Full given name is Richard Thomas Harold.

Alton, Florence and Harry. Parents of Dick Alton and biological paternal grandparents of Oliva.

Army.* Author's male springer spaniel in Delta, Utah.

Astor IV, John Jacob. Wealthy New Yorker who died on the Titanic

Aunt Mary. Author's maternal aunt, Beck, Mary Isadore.

Baker, Josephine. American-born black singer and dancer who became French citizen and adopted twelve children.

Bamberger, Simon. Jewish businessman with railroad and mining interests. In 1917 he was the first non-Mormon governor of Utah and remains the only Jew elected to that office. Also cousin of author's paternal great grandfather.

Bartelme, Kenwood Francis. Berkeley clinical psychologist who treated author.

Bear, Male.* Bear who broke into house on Nora Creek Road, near Troy, Idaho in March 1967.

Bear, Mother with Twins.* Bear who loved cherry chocolates of author's mother in Yellowstone in August 1950.

Bear, Mother with Twins.* Idaho bear who ate Tiger Cat on Nora Creek Road, near Troy, Idaho in November 1968.

Beautiful. Bart Penny's nickname for author's mother.

Beck, Edward Robert. Author's maternal grandfather, a.k.a. Pepa.

Beck, Edward Smoot. Author's maternal uncle.

Beck, Jack Campbell. Author's maternal uncle.

Beck, Margaret Audrey. Maiden name of author's mother, a.k.a. Mugs.

Beck, Mary Isadore. Author's maternal aunt, a.k.a. Aunt Mary.

Beck, Vermilla. Author's maternal aunt, born March 1, 1911 in Salt Lake City, Utah.

Bendel, Bill. Owner of Bendel's Meat Packing plant on Nora Creek Road, near Troy, Idaho.

Benson, Hermine. Olivia's kindergarten teacher in Troy, Idaho.

Berg, Norman. Owner of the Inland Hotel, Troy, Idaho.

Bird, Dr. Myron Evans. Physician in Delta, Utah in early 1940s.

Blackie.* Author's female cat in Logan in 1948-1949.

Blanton, Charles. Attorney who defended Keith Taylor in the 1967 hay truck accident trial.

Bosch, Hieronymus. Fifteenth century painter of "The Garden of Earthly Delights."

Boyd, Peter J., Esq. Attorney who represented author and Lary Carpenter in the 1967 hay truck accident trial.

Briggs, Susan. Author's classmate at EBHS in Bakersfield, who traveled to Europe with her in 1961.

Brocke, Frank. Owner of the First Bank of Troy, Troy, Idaho.

Brotman, Jordan. Owner of property on Nora Creek Road, near Troy, Idaho.

Brotman, Norah. Friend of author at UC Berkeley, who offered use of her property on Nora Creek Road.

Brown, Kim. Author's classmate at EBHS in Bakersfield, who traveled to Europe with her, a.k.a. Kim Federici.

Budge, Dr. Omar Sutton. Logan, Utah "X-ray specialist" who treated author and lead 1951-1952 "Walking Blood Bank" campaign for the Atomic Tattoo.

Budge, Elizabeth. Dated Teddy Wilson in Logan, Utah in 1956.

Calculus, Rebecca. Teenage daughter of Professor Calculus who helped care for Olivia in Berkeley, 1963-1965.

Calculus, Professor. Author's neighbor and Professor of Mathematics at UC Berkeley.

Carpenter, Betty Lewis. Mother of Lary Carpenter.

Carpenter, Lary. Classmate at EBHS and UC Berkeley who married author to avoid draft during Vietnam War. She divorced him in 1974.

Carpenter, Les. Father of Lary Carpenter; EBHS coach and advisor in Bakersfield.

Carr, Ronnie. Teenage passenger in 1966 hay truck accident.

Carvey, Dana. Comedian who plays "The Church Lady."

Cayce, Edgar. American clairvoyant and author.

Christie, Julie. American movie actress; in 1965 played Lara Antipova in the film version of Boris Pasternak's novel, Dr. Zhivago.

Cobin, Ellie. Sister-in-law of author, wife of Joel Pelcyger.

Cohen, Leonard. Canadian singer, songwriter, and poet.

Cole, Bev. Helped found Friends' School in Boulder, Colorado; mother of Cole Davis.

Cousin Julie Ann. Maternal first cousin of author and daughter of Aunt Mary.

Dahl, Robert Mahlen. Mormon missionary in France.

Davis, June. Student nurse who cared for author in Delta.

De Foucaucourt, Baronesse. Catholic woman who provided author with free lodging in her mansion on Boulevard St. Germaine in Paris, spring 1962.

Dooley, Dr. Tom. Physician and author who established MEDICO clinics in remote parts of Laos.

Downs, Lewis. Author's sixth grade teacher at Adams School.

Ekedal, Jim. Boyfriend and husband of author's best friend, Nonny Thomas. Full given name was James Michael Ekedal.

Erickson, Ronald. Classmate of author in Logan, Utah, 1955 who assisted with parallel parking of Chrysler.

Evans, Afton Lee. Substitute mother for author in Logan, Utah.

Evans, Ben. Father of Marlene, husband of Afton, and author's iconic model of equanimity.

Evans, Marlene. Author's best friend in Logan, Utah from 1947-1950, daughter of Afton and Ben Evans.

Fisher, Eddie. Hollywood movie star who left Debbie Reynolds for Elizabeth Taylor in 1958.

Fox, Dr. Maurice. Physician, who treated the author as Chief Resident at SF Medical Center in 1963.

Fromm-Reichmann, Frieda. Freudian psychotherapist who promoted cold, unloving mothers as the cause of schizophrenia, labeling those mothers as "schizophrenogenic."

Fuhriman, Jerry. Author's classmate, Logan, Utah, 1952-1954.

Funk, Alda. Author's neighbor in Logan, Utah, 1947.

Funk, Helene June. Author's homemaking teacher, Logan Junior High, Logan, Utah 1956.

Gannon, Michael I. American Jesuit, living in Paris in 1962, who helped author find volunteer placement in Africa.

Gilder, Glenn. Author's neighbor in Troy, Idaho 1967-1968.

Gilder, Agnes Clark. Author's neighbor and wife of Glenn Gilder.

Gottfredson, Faye and Ken. Members of author's parents' bridge club, Logan, Utah 1955.

Grandma Margaret. Olivia's maternal grandmother, Margaret Audrey Beck Lieberman.

Grandpa Frank. Olivia's maternal grandfather, Frank V. Lieberman.

Greaves, Verba. Mother of Vicky Greaves, author's classmate.

Greaves, Vicky. Author's classmate, Logan, Utah 1948-1950.

Grey Cat.* Olivia's female calico cat acquired from barn near Troy, Idaho in winter of 1967.

Hales, Mrs. Barbara M. Author's fourth grade teacher at Woodruff Hall, Logan, Utah, 1951-1952.

Hardenburgh, Dr. Firmon. Olivia's Harvard trained ophthalmologist.

Heavenly Father. Mormon nomenclature for God.

Hill, Mrs. Author's third grade teacher at the Ellis School, Logan, Utah in 1950.

Hinckley, Rulon T. President of French Mormon Mission, 1962.

Holmes, Judge Horace. Boulder County Juvenile District Court Judge who presided over Oliva's adoption by Bob Pelcyger, September 1975.

Hsu, Immanuel C. F. Professor of Chinese History at UC Santa Barbara, 1960.

Hutchins, Emmett. Early disciple of Ida Rolf who Rolfed author in Boulder in 1973.

Jean Maire, ZiZi. French singer and dancer, Paris, France, 1962.

Johnson, Sonia. Mormon woman excommunicated in 1979 for advocating for the Equal Rights Amendment.

Jones, Mary Sarah Ann. Author's maternal great grandmother, wife of William Cochran Adkinson Smoot.

Kerr, Clark. UC President during the Free Speech Movement; referred author to Jule Sugarman for .work with Project Head Start in 1968.

Kerr, Kay. Wife of UC President Clark Kerr.

Kissinger, Henry. Secretary of State under President Nixon.

Koiter, John. Dutch-born Mormon, owner of house rented by Lieberman family in Delta, Utah in 1942.

Kostof, Spiro. UC Berkeley Professor of Architectural History, born in Turkey; author translated research articles for him in Berkeley in 1966.

LaPointe, Oran. Oglala Sioux Indian who worked at NARF.

Larson, Royce. Author's classmate in Logan, Utah 1949-1951.

Leigh, Robert Jacob. Author's paternal uncle, who changed his surname from Liebermann to Leigh.

Liebermann, Alice. Author's paternal aunt, also known as Aunt Alice, who married Arthur William Marshall.

Liebermann, Francis Valentine. Name of author's father at time of his birth.

Liebermann, Jacob. Author's paternal grandfather.

Lieberman, Frank V. Author's father, born August 26, 1911, Newark, New Jersey.

Lindstrom, Mrs. Author's widowed landlady on Derby Street, Berkeley, California, 1962–1963.

Lurch. Author's descriptor for mortician at Howe Mortuary in Boulder in 1982.

Lusak, Stephanie. Lary Carpenter's first domestic partner with whom he had four children.

Jeanne Mere, Ma. French Mother Superior at Foyer de Bonte, Sarlat, Dordogne, France, 1962-1964.

Manx.* Son of Grey Cat, who was born with stubbed tail in Boulder 1970.

Marks, Charlotte Anne. Wife of David H. Getches and hostess phenom, also known as Ann M. Getches, ONINer and close friend of author.

Marquardt, Dr. James E. Boulder psychiatrist; treated author from 1973-1983.

Marshall, William Arthur. Author's first paternal cousin, son of Aunt Alice, also known as Bill Marshall.

Marshall, Arthur William. Husband of Alice Lieberman and author's paternal uncle by marriage.

McCarthy, Senator Eugene. Minnesota Senator; anti-war candidate for Democratic nomination for President in 1968.

Menninger, Dr. Karl. Psychiatrist; founder of the Menninger Clinic in Topeka, Kansas, who was the first to identify the link between in utero exposure to the influenza virus and mental illness.

Millie, the sow.* Devoted sow of Glenn Gilder.

Moore, Charles. UC Professor of Architecture, author's friend.

Moulding, Hockey. Friend and admirer of Anna Tuite Lieberman and her four children when they lived in Ogden, Utah. Given name was Harland.

Nanie. Author's maternal grandmother, see Vermilla Smoot Beck.

Nielson, Donna and Tom. Neighbors of Lieberman family in Logan, Utah, 1950-1956.

Nielson, Tana and Teri. Daughters of Donna and Tom Nielson.

Norrie.* Female cat, abandoned as kitten and rescued by author on beach near UC Santa Barbara.

Nurse.* Female springer spaniel, who was author's second dog in Delta, Utah 1942-1945.

Olly.* Bart Penny's rescued barn owl.

Olsen, Alice. Paternal grandmother of childhood friend and neighbor, Linda Olsen.

Olsen, Scott. Author's classmate in Logan, 1954-1956.

Olsen, Linda. Author's friend in Logan, 1948-1950.

Olivia. Author's daughter, see Pelcyger, Olivia.

Pablo.* Male Australian Sheep dog purchased by author and Lary Carpenter in Idaho, 1967.

Pelcyger, Bob. Author's husband of forty-two years, full given name is Robert Stuart.

Pelcyger, Ellie Cobin. Author's sister-in-law, wife of Joel M. Pelcyger, co-founder P.S. 1 in Santa Monica.

Pelcyger, Eugene. Author's father-in-law, father of Bob Pelcyger.

Pelcyger, Gale. Author's step-daughter.

Pelcyger, Joel M. Author's brother-in-law, husband of Ellie Cobin, co-founder P.S. 1 in Santa Monica.

Pelcyger, Jordan. Author's step-son, son of Bob Pelcyger and his first wife Deborah.

Pelcyger, Olivia. Author's daughter, mother of Tate and Eve.

Pelcyger, Ruth Cantor. Author's mother-in-law, wife of Eugene.

Penny, Bart. Author's boyfriend from Bozeman, Montana, 1957-1962.

Pepa. Author's maternal grandfather, see Beck, Edward Robert.

Piaf, Edith. French singer, Paris, France in 1962.

Pierce, Linda Louise. Second wife of Dick Alton, Olivia's biological father.

Rand, Ayn. Russian-born author of The Fountainhead who advocated for rational self-interest.

Rebout, Dr. Renee. French female physician in Ouagadougou, Haute Volta (now Burkino Faso), 1962.

Reid, Rose Marie. A Mormon woman and successful swimsuit maker, who worked to increase number of Jews converted to Mormonism.

Reynolds, Debbie. Hollywood movie actress who was betrayed by her husband, Eddie Fisher in 1958.

Ricks, Marc. Conducted excommunication trial of author from the Mormon Church in 1966.

Roark, Howard. Name of rebellious young architect in Ayn Rand's 1943 novel, The Fountainhead.

Rolf, Ida Pauline. German-born biochemist, developed deep tissue body work technique known as Rolfing.

Rosen, Dr. John Nathaniel. Author of "The Perverse Mother" whose false theories about the causes of and cure for schizophrenia haunted the author between 1959 and 1983. He was born Abraham Nathan Rosen.

Ryan, Mrs. Sallaee. American Field Services advisor and business teacher at EBHS in Bakersfield, California in 1959-1960.

Samuelson, Don William. Governor of Idaho, 1967-1971, who asked Idaho legislature to pass a proclamation "Kick a Beatnik in the Seat" week in 1967.

Savio, Mario. Led Free Speech Movement, Berkeley, 1964.

Schweitzer, Dr. Albert. French-German theologian, physician, author, Nobel Peace Prize winner, who operated a medical clinic in Gabon, Africa. He wrote "Out of My Life and Thought" a 1933 autobiography given to the author by Dr. Norris H. Weinberg in 1960.

Sharif, Omar. Egyptian actor who portrayed Yuri Zhivago in 1965 movie, Dr. Zhivago, based on the Boris Pasternak's novel of the same name.

Shore, Walter Evert. Author's biology teacher at EBHS in Bakersfield, California in 1958-1959.

Sigler, William F. Professor of fish ecology at Utah State University, employed author's mother in 1951-1952.

Skinner, Roger. Mormon missionary in France, who attended Gallatin County High School at the same time as the author in Bozeman, Montana.

Smith, Joseph. Mormon Prophet who "translated" Book of Mormon. Murdered by mob in June 1844.

Smith, Pastor James Comfort. St. John's Presbyterian Pastor; officiated at the marriage of the author and Dick Alton in Berkeley on March 23, 1963.

Smoot, Abraham Owen. Author's maternal third great grandfather, who was born in Kentucky in 1815 and converted to Mormonism in 1835. He died in Provo, Utah on March 6, 1895, having married five different women. The author is a descendant of his first wife, Margaret Thompson McMeans, who was a divorcee at the time of their marriage in 1838. He adopted her son from her first marriage, William Cochran Adkinson Smoot, who is author's maternal second great grandfather.

Smoot, William Cochran Adkinson. Adopted son of Abraham Owen Smoot and Margaret Thompson McMeans, born in Tennessee in 1828. Like his adopted father he practiced polygamy. The author is a descendant of him and his first wife, Martha Ann Sessions, whom he married in Salt Lake City in 1852. Their first child was a son born in 1853, also named, William Cochran Adkinson Smoot.

Smoot, William Cochran Adkinson II. He married Mary Ann Sarah Jones on March 28, 1858 in Salt Lake City and they had five children. Their youngest child was the author's maternal grandmother, Vermilla Smoot, also known as Nanie.

Smoot, Vermilla. Author's maternal grandmother, also known as Nanie Smoot Beck, born October 30, 1889 and died March 28, 1954.

Snow, Mrs. Meda. Author's piano teacher in Logan, Utah 1954.

Snowball.* Female cat belonging to author in Logan, Utah in 1948-50.

Spettigue, Beulah. Niece of Nanie Smoot who died of Spanish Influenza in 1919 at age ten.

Spettigue, Genevieve Smoot. Mother of Beulah; sister of Nanie (Vermilla) Smoot Beck.

Spotty.* Olivia's male kitten lost in Oregon hay truck accident.

Sprague, Margaret Lucius. Author's geometry teacher at EBHS, in Bakersfield, California, 1958-1959.

Stripp, Fred Sheridan. Professor of Rhetoric at UC Berkeley who officiated at the marriage of the author and Lary Carpenter in Berkeley, March 27, 1965.

Stripey.* Olivia's female kitten lost in Oregon hay truck accident.

Sullivan, John M. Friend of the author and graduate student in political science at University of Idaho, who helped establish the New Party and worked for President Jim Dixon at Antioch College. John, his wife Patricia, and their four children were living in Moscow, Idaho in 1967-1968.

Swift, Porky. Fellow passenger on SS Groote Beer in September 1961; one of the Spokane boys who became a friend of the author. His given name was Paul Swift.

Taner, Ayse. Turkish American Field Service exchange student who lived with the author's family in Bakersfield, California in 1959.

Tante Martha. Impoverished elderly aristocrat, living at Foyer de Bonte, who tutored author in French, 1962.

Taylor, Don. Father of teenage driver; owner of hay truck in 1966 Oregon accident.

Taylor, Elizabeth. Famous Hollywood actress who had seven husbands, one of whom was Eddie Fisher.

Taylor, Keith. Teenage driver, 1966 hay truck accident.

Teda. Author's lovey, a handmade teddy bear created by her maternal grandmother, Nanie.

Temple, Shirley. Famous Hollywood child actress in the 1950s.

Thomas, Nonny. Classmate and best friend of author at EBHS and beyond. Given name at birth was Naneen Francis Thomas.

Tiger Cat.* Grey Cat's litter mate in Idaho who had a stub tail and was eaten by bears in house on Nora Creek Road in November 1968.

Tiger Two.* Female kitten, Grey Cat's only Colorado litter.

Tuite, Anna Gertrude. Maiden name of author's paternal grandmother. Also known as Anna T. Liebermann and as Nancy Liebermann.

Turvey, Dr. Edward. Boulder cardiologist for author's father and mother in Boulder, Colorado, 1975-1993.

Uncle Bob. Author's paternal uncle, Robert Jacob Liebermann, who changed his surname to Leigh after harassment during World War II.

Van der Ryn, Sim. Professor of Architecture, UC Berkeley in 1960s, friend of author.

Verhaag, Mildred. Manager of Bridal Salon, Brock's Department Store in Bakersfield, 1958 – 1963.

Wallway, Beatrice Gilfert. Friend of Lieberman family who cared for author in Delta, wife of Marvin.

Wallway, Marvin. Specialist sent to Delta by War Department in 1942 to search for uranium; shared office space with author's father and became close friend of Lieberman family.

Wallway, Mary Kay. Author's first friend in Delta, 1942- 1945. Full given name was Mary Katherine.

Watkins, Sherry. Author's friend in Logan, Utah 1950-1956.

Weinberg, Dr. Norris H. UCLA Psychologist that author consulted in 1960 about her mother's schizophrenia.

Wickel, Samuel R. Owner of Wickel's Men's Store in Logan, Utah, 1954.

Wilson, Teddy. Author's boyfriend in Logan in 1955-1956; his given name was Theodore John.

Winnicott, Donald Woods. British pediatrician and psychoanalyst who wrote extensively about the role of mothers in the psychological development of their children. Also known as D.W. Winnicott.

Wolf, Dr. George R. Physician at Caldwell Memorial Hospital who treated Olivia, Lary Carpenter, and author for injuries suffered in hay truck accident in 1966.

Woods, Sherry LaTrelle. EBHS classmate in Bakersfield, who traveled to Europe with the author in fall of 1961.

Work Mistress. Name of symbolic persona for Bob Pelcyger's professional work demands.

Young, Arlene Garber. Boulder Head Start teacher and mother of Zoe, Olivia's best friend at Upland School in 1969-1971.

Young, Brigham. Second prophet of Mormon Church; he succeeded Joseph Smith after Smith was murdered in Carthage, Missouri.

Young, Elaine. Neighbor and life sciences teacher who lived in basement apartment of rental house in Logan, Utah, 1948-1950.

Young, Stan. Husband of Elaine; neighbor in basement apartment at 435 on 500 West in Logan, 1948-1950, student at Utah State University.

Young, Steven. Professor of neuropsychology at CU Boulder and father of Zoe, Olivia's friend.

Young, Zoe. Olivia's best friend at Upland School in Boulder, 1969-1971.

Credits for Photographs and Illustrations

All photographs and illustrations in Part One of *Optimal Distance, A Divided Life*, are from the collections of the Beck-Lieberman family or the Cantor-Pelcyger family, unless otherwise noted below.

Page

25 Topaz Relocation Center near Delta, Utah, 1943.
Reprinted with the permission of the J. Willard Marriot Library, University of Utah, digital collection, Topaz Relocation Center.

28 Topaz Relocation Center at Night, 1945.
Oil on Canvas by George Matsusaburo Hibi.
Reprinted with permission of Department of Special Collections, University Research Library, University of California at Los Angeles.

56 Baptismal Font in Logan LDS Temple, about 1947.
Reprinted with the permission of Utah State University, Harry Reuben Reynolds Photograph Collection.

79 Illustration for placement of blood type tattoo, August 1, 1950.
Reprinted with the permission of the *Chicago Tribune.*

83 Wickel's Men's Store, 81 North Main Street, Logan, Utah, 1955.
Photograph by H.S. Crocker Company; Picture ID: 517
Reprinted with the permission of the Logan Public Library – from the John F. Robinson Historic Photo Collection.

299 The author and Olivia at home of David H. and Ann Marks Getches, on Twin Sisters Road, Boulder County, November 1971.
Photograph by Dan Budnik.

321 Olivia as a student at Boulder High School, Boulder, Colorado, 1978.
Photograph by Judith Mohling.

Endnotes

1. Zions Cooperative Mercantile Institution (ZCMI) was founded by Brigham Young in 1868 with the support of a vote from the Council of Fifty, one of the early layers of male LDS decision-making.

2. DDT (dichlorodiphenyltrichloroethane) was first synthesized in 1874. In 1939 Paul Hermann Muller, a Swiss chemist, discovered the insecticidal actions of DDT, a discovery for which he was subsequently awarded the Nobel Prize in Medicine. The U.S. Army began using DDT during the second half of World War II to control malaria, typhus, lice, and bed bugs among troops. The USDA simultaneously assigned Frank V. Lieberman to test DDT for agricultural use. After the war, DDT was made available as an insecticide. The use of DDT was banned in 1972, ten years after Rachel Carson wrote Silent Spring. (Carson, Rachel. *Silent Spring*. New York: Houghton Miffin, 1962.)

3. Although George Rieveschi, an Ohio chemical engineer, received most of the credit for the development of Benadryl, it was the Swiss-born Italian pharmacologist, Daniel Bovet, who first created its predecessor compound, (neoantergan which produced severe drowsiness), for which Bovet received the Nobel Prize in 1957. According to Beck family oral history, a Mormon medical student serving a mission in Switzerland, who suffered from severe allergies, brought a supply of neoantergan back to Utah. Margaret was given some of his supply when she was hospitalized in Salt Lake City. She continued to treat her allergies with it and then with Benadryl for many years, likely suppressing symptoms of her biological mental illness.

4. Temple Garments, also known as "Mormon underwear" (or by some comedians as "magic underwear"), are items worn beneath clothing by faithful members of the Church of Jesus Christ of Latter Day Saints to remind them of the sacred commitments they have made in various Temple ceremonies. Such garments serve a purpose similar to those worn by members of other religious groups such as the cassocks of Catholic priests, the Jewish prayer shawl, the Muslim's skullcap, or the saffron robes of Buddhist monks.

5. The Relief Society was organized by Joseph Smith on March 17, 1842 "to look after the spiritual welfare and salvation of all the female members of the Church." Operating under the direction of Mormon priesthood leaders, it has multiple purposes, including preparing women for the blessings of eternal life by increasing their faith in Jesus Christ, strengthening families and homes through teaching women ordinances and covenants, and encouraging women to work together to help those in need.

6. The Bamberger Electric Railroad, known as "The Bamberger" began operating in 1891. Built by Simon Bamberger, it provided both passenger and freight service between Ogden and Salt Lake with stops in between, including the Lagoon Resort. Passenger operations ended in 1952.

7. A Mormon Ward House is the same as a church. Mormon communities are divided into wards and members attend the ward to which they are assigned, usually by the location of their residence. Each Ward has a bishop, an unpaid leadership position held by a member who lives in the area and is considered to have outstanding character.

8. Professor Raymond J. Becraft created the garden on the property purchased by the author's parents. Born March 16, 1890 in Ogden, Becraft graduated from Utah State University in 1917, where he was elected student body president. He married Ireta Harris December 23, 1919 in the Logan Temple and became a Professor of Range Management at Utah State. He and Ireta bought a home at 606 North 400 East, as well as the adjoining lot at 624 North 400 East, for his experimental garden. In 1935 Professor Becraft joined the faculty of the University of Idaho in Moscow, Idaho. In 1937, Becraft was appointed Regional Director of Plant Ecology for the U.S. Forest Service in Portland, Oregon. It was in Portland on September 6, 1938, at age forty-eight, Professor Raymond Becraft shot and killed his wife and seventeen-year-old son, Raymond Harris, before taking his own life. Survivors included three daughters, then sixteen, ten, and four. Newspapers from that era indicate that the murders-suicide were notorious for many years. I do not know whether my parents knew of the circumstances surrounding the death of the man who created the backyard garden of the Dream House, but I now wonder whether knowledge of the Becraft family tragedy triggered the hallucinations of the author's mother about the ghost in the garden.

9. Professor Sigler, head of the Utah State University Department of Wildlife Management, attended Iowa States University during the same period of time as Frank V. Lieberman. Dr. Sigler hired Margaret Audrey Beck Lieberman to work as his

stenographer at Utah State University in 1952-1953. See Sigler in the Biographical Index at www.OptimalDistance.com/Information for information on subsequent synchronistic link to Lieberman family.

10. The GI Bill was the informal name for the Servicemen's Readjustment Act of 1944, a law providing a range of benefits for returning World War II veterans, including cash payments of tuition and living expenses to attend a college or university.

11. "Atomic Tattoo" was a 1950s Civil Defense program more commonly referred to as a "Walking Blood Bank." Implemented as an experimental program in Cache and Rich Counties in Utah and a few counties in Indiana, Dr. Omar Sutton Budge was a forceful advocate for the program in Logan. On August 18, 1950, *The Logan Herald Tribune* reported on a civic meeting about the atomic tattoo program:

> *"Dr. Omar Budge, chairman of the committee in charge of the blood typing program for Cache County, explained the life-saving advantages of having every person's blood tested and the type of his blood recorded on the person's skin by a simple tattooing process."*

Also see: "Many Utahns Bear Reminder of Cold War" *Salt Lake Tribune*, August 14, 2006 and www.conelrad.com/atomicsecrets.

12. Senator Joseph Raymond McCarthy of Wisconsin came to embody the fear of Communist subversion during the Cold War, perpetuating the kind of fear mongering that ruined careers and lives. Although eventually censured by the Senate, for many years Joseph McCarthy fed a mixture of national suspicion, slander, and paranoia that came to be known as "McCarthyism." Senator McCarthy represented Wisconsin from 1947-1957.

13. In 1956, any Montana citizen who was fourteen years of age or older was eligible for a driver's license.

14. For a fuller explanation of this phenomena see: Rosabeth Moss Kanter with Barry A. Stein. *A Tale of "O" - On Being Different in an Organization.* New York: Harper Colophon, 1980.

15. Dieldrin was developed in 1948 as an alternative to DDT. It was used extensively between 1950 and the early 1970s. Unfortunately, it is an extremely persistent pollutant; long term exposure has shown toxic side effects in animals and humans. Dieldrin has been linked to Parkinson's, breast cancer, and other immune, reproductive, and nervous system

damage. Because it has been shown to be an endocrine disruptor, Dieldrin is now banned in most of the world.

16. The Quorum of Twelve Apostles (also known as the Quorum of the Twelve, the Council of the Twelve Apostles, or simply The Twelve), first organized in 1835, is the second-highest governing body in the Mormon Church hierarchy. All members are men.

17. The American Field Service (AFS) began as the American Ambulance Field Service in 1915. It was founded by A. Piatt Andrew, a former assistant professor of economics at Harvard, as an international non-profit organization to provide students with the knowledge, skills, and understanding needed to create a more just and peaceful world.

18. Frieda Fromm-Reichmann, (1889-1957) was a German-born psychoanalyst, who was a contemporary of Sigmund Freud. Dr. Fromm-Reichmann believed: *"The schizophrenic is painfully distrustful and resentful of other people, due to the severe early warp and rejection he encountered in important people of his infancy and childhood, as a rule, mainly in a schizophrenogenic mother."* Fromm-Reichmann, F. "Notes on the development of the treatment of schizophrenics by psychoanalytic psychotherapy." *Psychiatry.*1948; 11(3):263-273.

19. "The Perverse Mother" by Dr. John Nathaniel Rosen (1901-1993) was the most damning article read by the author in December 1959. The words of Dr. Rosen haunted her for years. In it he said, "A schizophrenic is always one who is reared by a woman who suffers from a perversion of the maternal instinct." Rosen, John N., (1953) *Direct Analysis* (selected papers), (pp. 97-105), Grune and Stratton, New York City, NY.

20. Dr. Karl A. Menninger (1893-1990), a Harvard trained psychiatrist, who founded the Menninger Clinic in Topeka, Kansas, was among the first to identify influenza as a risk for biological mental illness. Menninger, Karl A., "An Analysis of Post-Influenzal Dementia Precoxas of 1918, and Five Years Later," *Psychiatry,* Vol. 82, Issue 4, April 1926: 469-529.

21. Albert Schweitzer, born 1875 in Alsace, France, was a theologian, philosopher, and physician. One of his numerous books was given to the author in 1960 by Dr. Norris H. Weinberg. Schweitzer, Albert. *Out of My Life and Thought* New York: Henry Holt, 1933. Dr. Schweitzer died in Lambaréné, Gabon in 1965.

22. After Jacob Liebermann abandoned his wife and four children, Anna Tuite Lieberman was without a source of income. She was forced to place the author's father, his two brothers, and his sister Alice in the Martha Society Home in Ogden, Utah.

Directories for Ogden show Anna took a room in a boarding house while working at "Kennedy's Cafeteria." A news story published in the *Ogden Standard-Examiner* described the Martha Society Home as follows: *"For years the Martha Society home has been a help and a standby to parents, who are unable to property care for their children. With the idea that the parents should still retain some of the responsibility, a small fee is asked by the society for every child and the remaining expense is paid by the society. Clean food, clean clothing, and clean entertainment is furnished for the little folks."* *Ogden Standard-Examiner*, Sunday, October 30, 1921, page 25.

23. Susan Briggs and Sherry Wood returned to the United States as scheduled at the end of December 1961. Kim Brown returned to America in January 1962. Paul Swift decided to continue traveling, rather than to volunteer in Africa and returned in December 1962.

24. Josephine Baker was born Freda Josephine McDonald in St. Louis, Missouri on June 3, 1906. Her career as a black singer and dancer made her famous in France, where she became known as "The Black Pearl;" she subsequently became a citizen of France. Josephine was married four times; the first marriage took place when she was only thirteen. After being active in the civil rights movement, she dreamt of a multi-cultural and multi-racial family, eventually adopting a dozen children from different countries. The author met Josephine, her fourth husband, Jo Bouillon, and several of her children when they were living in Chateau des Milandes, Dordogne, France.

25. Chlorpromazine, brand name Thorazine, was the first drug developed that had specific anti-psychotic action and was used to treat schizophrenia. It possesses antiadrenergic, anti-serotonergic, anticholinergic, and antihistaminergic properties. Synthesized in France in 1951, it was licensed in America to Smith Kline & French (today GlaxoSmithKline) in 1953. While physicians were slow to adopt it as a treatment, the impact of Thorazine in emptying psychiatric hospitals has been compared to that of penicillin on infectious diseases. It slowly replaced electroconvulsive therapy, hydrotherapy, psychosurgery, and insulin shock therapy.

26. Like most patients suffering from schizophrenia, the author's mother heard threatening voices telling her what to do. After her initial hospitalization in Tucson, those voices returned whenever she wasn't taking antipsychotic medications.

27. The author still has the pamphlet entitled *"Some Day You Will Marry"* in her possession. It was written by Angelyn W. Wadley and published by the Mormon Church's Young Women's Mutual Improvement Association.

28. A small leaflet published by the Mormon Church and used in their Mutual Improvement Association (MIA) teenage education programs.

29. The Campanile, or what is also known as Sather Tower on the campus of the UC Berkeley, is a bell and clock tower that is the third tallest such tower in the world. It is the central landmark of the Berkeley campus.

30. Ayn Rand (born Alisa Zinovyevna Rosenbaum in St. Petersburg, Russian in 1905) immigrated to the United States in 1926. She developed a philosophical system called "objectivism" based on rational self-interest and self-responsibility. Rand first achieved fame with her novel, *The Fountainhead* published in 1943. Her best known work was *Atlas Shrugged*, published in 1957. She believed in rational and ethical egoism, and rejected altruism.

31. Sonia Ann Harris Johnson, the leader of "Mormons for ERA" (Equal Rights Amendment) was excommunicated in December 1979 after having been charged by Mormon authorities with a variety of misdeeds, including hindering the worldwide missionary program, damaging internal Mormon social programs, and teaching false doctrine.

32. *The Berkeley Barb* was a weekly underground newspaper published between 1965 and 1980. For many years it provided more detailed information about the Vietnam War and the Free Speech Movement than the standard press.

33. Summerhill was an English school founded by A. S. Neill (1883 - 1973) in 1924 in Lyme Regis. In 1927, the school moved to Leiston, where the curriculum continued to be focused on the philosophies of freedom from adult coercion and community self-governance. Summerhill was well-known in the 1960s and 1970s due to interest from the counterculture movement and from A.S. Neill's book. Neill, A.S., *Summerhill, A Radical Approach to Child Rearing*, (London: Hart Publishing, 1960).

34. The Glide Foundation is a Methodist sponsored charity, whose mission was to break the cycles of multi-generational dependency, poverty, and low self-worth by providing a spiritual home of unconditional love.

35. Upland School was founded by Iris and John Green in 1967. It closed in June 1988. For many years, the author served on Upland School's board of directors.

36. Edgar Cayce (1877 - 1945) was an American clairvoyant and psychic, who used a trance or hypnotic state to predict future events.

37. Lary and Stephanie had four children—three daughters and a son. Their twelve years of common law marriage ended in 1981. He then began a relationship with a nearby neighbor, "Wind," with whom he had two children. Their sixteen-year-old son committed suicide on March 1, 1998. Lary's fourth partner or common law wife was known as "Bonnie Blackberry." They had no children together. In the last decade of his life, Lary took up watercolor painting and continued to maintain his large garden. Lary Carpenter died of lymphoma in Humboldt County on August 6, 2005. He was sixty-three.

38. Norman O. Brown (1913-2002), joined the faculty at the University of Rochester in 1962, Bob's senior year. In 1959 Brown had written the book for which he is best known, *Life Against Death—The Psychoanalytical Meaning of History*. In Brown's effort to understand world history through psychoanalysis, he raised profound and stimulating questions. Bob Pelcyger was one of eight students who participated in Brown's seminar during spring semester 1963. The seminar opened Bob's mind for the first time to the mysteries of the unconscious, on both an individual and collective level. It also introduced him to poets like William Blake and to artists like Hieronymus Bosch. Bob often said that his exposure to Brown, to *Life Against Death* and to Brown's two subsequent books, *Love's Body*, (Random House, 1966) and *Closing Time*, (Random House, 1973) were essential components in the intellectual and emotional bridge that brought he and Joan Carol Lieberman together.

39. Norman O. Brown introduced Bob to the writings of Owen Barfield (1898-1997), a British philosopher and author of various works on language, myth, perception, and the evolution of human consciousness. Barfield was also a practicing solicitor until his retirement in 1959. After his legal career, he taught and lectured at various North American colleges and universities including, Brandeis, Drew, and SUNY-Stoney Brook. In 1957, he published *Saving the Appearances – A Study in Idolatry*, his major work on the history and evolution of consciousness. During Bob's Fulbright study in London, he and Barfield explored the history of various legal concepts, such as the insanity defense, for the light those concepts shed on how the relationship between human beings and the world they perceive has changed over time.

40. Ida Rolf (1896 - 1979) developed "Structural Integration" a system of manual therapy meant to improve functioning of the human body in relation to gravity, now known by the trademark "Rolfing." Her 1920 doctoral degree in biological chemistry was from Columbia University's College of Physicians and Surgeons.

41. The term "Refrigerator Mother" was first used by Leo Kanner in 1949 in a paper describing the parents of schizophrenic and autistic children. Such children were "exposed from the beginning to parental coldness, obsessiveness, and a mechanical type of attention to material needs only. They were left neatly in refrigerators which did not defrost."

42. Arlie Russell Hochschild, a UC Berkeley sociologist, brought her eloquent voice to this phenomena when she wrote about the lives of people working in an Ohio manufacturing plant in her 1989 book, *Second Shift* (New York, Viking, 1989). Work provided the safest, most satisfying relationships for Hochschild's subjects because the rules of engagement were spelled out. Work rules and codes of behavior carefully defined the space between self and others, giving it more predictability and satisfaction than work at home. This was particularly true for women who faced "second shifts" when they returned home from work.

About The Author

In many ways, the roots of OPTIMAL DISTANCE — *A Divided Life*, extend back to June 17, 1947, the day after my fifth birthday. As I lay on my stomach on a surplus Army blanket in Logan, Utah, I used a crayon to illustrate my feelings in a pink diary with a praying child printed on the cover. Unable to spell more than a few words, I tried to draw how I felt. The diary had arrived that day in the mail, a birthday present sent by my Aunt Mary from Salt Lake City. It had a pseudo lock and key, but I soon lost the tiny aluminum key and had to break the lock. When the pink diary had no more empty pages, I began using a school notebook, a habit that persisted for the next seven decades. Writing has always kept me alive.

I was born in Utah in 1942. My father was a scientist and atheist, distantly related to Simon Bamberger, the first and only Jewish governor of Utah. My mother, a descendant of prominent Mormon pioneers, lost her familial faith during the Depression. Tragically, she developed paranoid schizophrenia shortly after my birth and, from then until her death, her mind was under the control of invisible demons. As an only child in Logan, Utah, I took refuge in the interstices of the Mormon Church trying to keep a safe distance from my mother's unpredictable and murderous impulses because I knew she would never follow me there. I often felt like a small wild animal desperately hiding from danger among a large herd of domineering dairy cows.

I was fourteen when I left Utah and Mormonism behind after my father was transferred to Bozeman, Montana. A year later he was sent to Bakersfield, California where I finished high school. Following a year of study at the University of California, I traveled in Europe and worked as a medical volunteer in Africa. When I returned to Berkeley, I gave birth to my daughter, pushing her stroller through demonstrations for Free Speech and against the War in Vietnam. In 1966, I left Berkeley to finish my thesis on leadership in a rent-free house in Northern Idaho. Two years later, a job offer to become the director of the county-wide Head Start program brought me to Boulder, Colorado, where I still reside. In 1971, I undertook an assignment for the Ford Foundation to help develop the Native American Rights Fund, where I met and fell in love with Bob Pelcyger. We were married in 1975, and for the next four decades, I worked as a management consultant serving clients who were lawyers, doctors, and women in leadership positions.

In 1999, as a finalist for the Bakeless Literary Prize, I was invited to attend the Bread Loaf Writers' Conference. My publishing consult was with Carol Houck Smith of W.W. Norton, who encouraged me to expand my workshop submission into an autobiography. Then fifty-seven, I had been "living" with metastatic cancer for a decade, and didn't believe I had enough time left on my life clock to undertake such an effort. Nonetheless, buttressed by Carol Houck Smith's endorsement, I began searching through my diary entries for clues and areas of research. Progress slowed after I suffered a near fatal stroke while flying to Utah in 2007. Sadly, Carol Houck Smith died the day after Thanksgiving in 2008, while improbably, I lived on.

After forty-two years of marriage, Bob and I made a bucket list for what each of us wanted from the other before our approaching deaths. The only item on Bob's list was the completion of OPTIMAL DISTANCE — *A Divided Life*. Now seventy-five and dependent on a life-sustaining cocktail of medications, I have finally fulfilled my beloved husband's request. Credit should go to Bob for his endless persistence and to Aunt Mary for the gift of a pink diary.